Numerical and Experimental Studies on Combustion Engines and Vehicles

Edited by Paweł Woś and Mirosław Jakubowski

Published in London, United Kingdom

IntechOpen

Supporting open minds since 2005

Numerical and Experimental Studies on Combustion Engines and Vehicles
http://dx.doi.org/10.5772/intechopen.87852
Edited by Paweł Woś and Mirosław Jakubowski

Contributors
Mohamed Ali Mergheni, Semakula Maroa, Freddie Liswaniso Inambao, Enhua Wang, Chenyao Wang, Fujun Zhang, Huasheng Cui, Chuncun Yu, Bolan Liu, Zhenfeng Zhao, Changlu Zhao, S.K. Gugulothu, B. Bhasker, V. V. Phani Babu, Paweł Woś, Krzysztof Balawender, Mirosław Jakubowski, Artur Jaworski, Paulina Szymczuk, Adam Ustrzycki, Simeon Iliev, Gonzalo Moreno Contreras, Daniel Martins, Rodrigo de Souza Vieira, Mohamed Mahdi Belhajbrahim, Toufik Boushaki, Jean-Charles Sautet

First published in London, United Kingdom, 2020 by IntechOpen
IntechOpen is the global imprint of INTECHOPEN LIMITED, registered in England and Wales, registration number: 11086078, 5 Princes Gate Court, London, SW7 2QJ, United Kingdom
Printed in Croatia

British Library Cataloguing-in-Publication Data
A catalogue record for this book is available from the British Library

Additional hard and PDF copies can be obtained from orders@intechopen.com

Numerical and Experimental Studies on Combustion Engines and Vehicles
Edited by Paweł Woś and Mirosław Jakubowski
p. cm.
Print ISBN 978-1-83968-032-8
Online ISBN 978-1-83968-033-5
eBook (PDF) ISBN 978-1-83968-034-2

For EU product safety concerns:
IN TECH d.o.o., Prolaz Marije Krucifikse Kozulić 3, 51000 Rijeka, Croatia, info@intechopen.com or visit our website at intechopen.com.

Meet the editors

Paweł Woś, PhD, DSc, associate professor at Rzeszów University of Technology, graduated in Mechanical Engineering from Rzeszów University of Technology in 1995. He earned his PhD degree in Internal Combustion Engine engineering in 2003. From 2005 to 2006, he served as a post-doc fellow at the Wayne State University in Detroit (USA) where he was involved in advanced combustion processes in ICE. Since then he continues these studies with application of variable compression ratio (VCR) technology in combustion engines design. He serves as an Associate Professor and Head of the Department of Automotive Vehicles and Transportation Engineering in Rzeszów University of Technology, Faculty of Mechanical Engineering and Aeronautics.

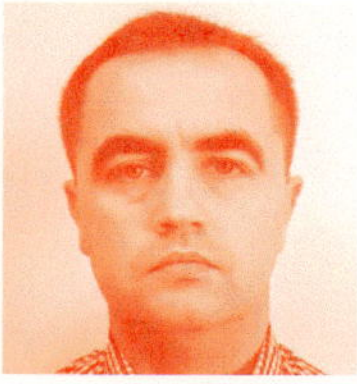

Mirosław Jakubowski, PhD, assistant professor at Rzeszów University of Technology, graduated in Mechanics and Machine Design from Rzeszów University of Technology in 2003. In 2006, he completed a six-month professional internship at MSX International Engineering GmbH in Munich, Germany (mpx Entwicklung GmbH Pininfarina Gruppe). He has been trained in design team specializing in automotive components and bodies. From 2008 to 2009, he worked as a CAD/Catia trainer at Leopard Automobile and ASSISTO. From 2005 to 2009, he continued doctoral studies on Construction and Operation of Machines at Rzeszow University of Technology, Faculty of Mechanical Engineering and Aeronautics. In 2012 he earned a PhD degree in Mechanical Engineering and serves as an Assistant Professor. His main interests are automotive technology and the use of modern technologies in machine design - computer aided design and reverse engineering technologies.

Contents

Preface

Computational numerical analysis and experimental research are the main tools used in many engineering disciplines, including the issues related to internal combustion engines and vehicle construction. Experimental studies are often associated with high cost and time-consuming works, therefore simulations and computational analyses are commonly used for performing parametric studies. Considering especially internal combustion engines, computational tests enable a detailed analysis of complex working processes taking place inside the engine. Moreover, they allow for shortening the time of testing numerous parameters and analysis of their influence on the final test result. They are also a reference point for the results obtained experimentally. The combination of computational methods with experimental tests undoubtedly opens up wide opportunities for scientists to improve existing and develop new technical solutions.

The matters discussed and presented in the chapters of this book cover a wide spectrum of topics and research methods commonly used in the field of engine combustion technology and vehicle functional systems. The book contains the results of both computational analyses and experimental studies on jet and reciprocating combustion engines as well heavy-duty onroad vehicles. Special attention is devoted to research and measures toward preventing the emission of harmful exhaust components, reducing fuel consumption or using unconventional methods of engine fueling or using renewable and alternative fuels in different applications. Some technical improvements in design and control of vehicle systems are also presented.

Sharing the book with our readers worldwide, the editors wish to thank all contributing authors and IntechOpen publishing initiative. We are especially grateful to Mrs. Mia Vulović, Mr. Luka Cvjetković, and all the publisher's staff coordinating the project; technical editors, and servicemen processing the book publication. Also many thanks to Ms. Paulina Szymczuk, MSc, for her great effort in book preparation.

Paweł Woś and Mirosław Jakubowski
Rzeszów University of Technology,
Rzeszów, Poland

Section 1

Jet Engine Modeling and Analysis

Chapter 1

Numerical Investigation of the Shock Train in a Scramjet with the Effects of Back-Pressure and Divergent Angles

Santhosh Kumar Gugulothu, B. Bhaskar and V.V. Phani Babu

Abstract

Numerical simulations are carried out to study the effect of divergence angle and adverse pressure gradient on the movement of shock wave train in a scramjet isolator. The commercial software tool ANSYS Fluent 16 was used to simplify two dimensional Reynolds averaged Navier Stokes equation with compressible fluid flow by considering the density-based solver with standard K-ε turbulence model. The species transport model with single step volumetric reaction mechanism is employed. Initially, the simulated results are validated with experimental results available in open literature. The obtained results show that the variation of the divergence angle and back pressure on the scramjet isolator has greater significance on the flow field. Also, with an increase in the back pressure, due to the intense turbulent combustion, the shock wave train developed should expand along the length and also moves towards the leading edge of the isolator leading to rapid rise in the pressure so that the pressure at the entrance of the isolator can match the enhanced back pressures.

Keywords: CFD, scramjet, isolator, divergent angles, back pressure and combustion

1. Introduction

The hypersonic air-breathing jet engines are designed to operate in the supersonic-combustion ramjet engine when the Mach number is more than 6. A scramjet engine incorporates with isolator, combustor, and nozzle. Isolator in the scramjet engine is widely used since its inception. Isolator in the scramjet engines is a typical component since it has a significant effect on the dual-mode engine transition. The hypersonic vehicle operates at a particular time during the ascent phase [1]. These effects happen due to the absence of mechanical compressor, free stream air as well as compression ratio. The propulsion vehicle maintained the engine by using inlet and isolator. The main task of the isolator is to differentiate the combustion pressure that occurs in the combustion chamber and should not reach the inlet [2].

Boundary layer interaction observed in the isolator when the dual-mode engine runs with ramjet engine. The air, pressure movement in an isolator, as well as shock train, are maintained the position of the Mach wave train. The compressed air is adjusted in isolator to match with the condition that can enter into the combustor. When the pressure introduced in the reverse direction of the combustor zone obtained the variation in pressure from the isolator and combustor. The difference between isolator and combustor zone is adjusted by changing the shock wavelength [3]. While designing the isolator, need serious concern about the unstart phenomenon. The isolator may lead to severe effect due to high speed in the flight. The length of the isolator part in the scramjet engine maintains at a certain weight. The required shear and shock waves provided to avoid the communication of the instabilities that will arise and affect the inlet [4]. The system developed using the hypersonic inlet isolator under Mach 4 and Mach 5 flight conditions [5]. In [6], reported that the decrease in the pressure at the inlet of the domain is observed with an increase in the isolator length. The shock train in a fixed 2-D scramjet inlet with isolator showed some results by increasing wall and decreasing the total temperature [7]. In Mach 5 inlet-isolator model, the shock train jumping moments captured by separating flow at the head of the shock train and the contraction ratio of the local throat-like shape [8]. The scramjet isolator decreases the static pressure, and it becomes sharper. The experiment conducted on a constant-area scramjet isolator and observed that was relatively stable with time-resolved and low-frequency pressure [9]. In [10], the numerical simulation influences the movement of free stream characteristics leading to separation with an increase in adverse pressure. The dynamic model of the shock train is predicted on the shock wave layer. The dynamic model cannot suppress the pressure gradient as high as the other sustains [11]. In [12], the complex compression and expansion waves exist in the isolator, causing large stream-wise and transverse gradients upstream of the shock train. The adverse gradient pressure in stream-wise decreases with the duct curvature [13]. In [14], the experiments compared with the conventional approaches using boundary layer interaction large-eddy simulation of a hypersonic of Mach 8 flight vehicle. In [15, 16], they have conducted experiments on the multiple shock wave/turbulent boundary layer interactions in a rectangular duct using Mach numbers 2.45 and 1.6. Carroll et al. observed that the length of the communications and the tendency towards a repeated oblique was scaled directly with the level of confinement. The study of unstart and unstarted flows in an inlet/isolator model strongly associated with boundary-layer separation [17]. The numerical solutions of the Naiver-stokes equations for the interactions of a shock wave and turbulent boundary layer varying from 7.93 to 12.17, at a free-stream Mach number of 2.96 and Reynolds number 1.2×107. The free-stream predicts accurate results. When shock strength and overall rise pressure for the low viscosity pressure asymptotes. The large-eddy 3-D analysis in the area of uniform cross section with low aspect ratio rectangular duct geometry is studied [18–22].

In the open literature, by varying the adverse pressure gradient at the exit of the isolator the motion path and characteristics of the shock wave train are obtained. In this work, we intend to study the impact of combustion phenomena on shock wave train. Therefore, the significance of the angle of attack on the wall surface of the domain and the effect of the adverse pressure gradient on the movement of the shock wave train are analyzed. The present analysis focused on different divergent angles, i.e., 0, 0.5, 1, and 1.5° with constant pressure gradient of 90 kPa, and also with different negative pressures of 80 and 100 kPa with constant cross-sectional area of the isolator is discussed. All these effects are studied on similar computational domain with similar solver type parameters. The rest of the paper is as organized as follows, Physical model and simulation methodology is discussed in Section 2, the effects of back pressure and angle of attack are discussed in Section 3.

2. Physical model, simulation methodology and validation of computational fluid dynamics code

2.1 Physical model

In this work, analysis has been carried out on the scramjet isolator of uniform cross section to analyze the movement of shock wave train to study the relation between shock train and the interaction of the boundary layer formation. As shown in **Figure 1** to study the impact of adverse pressure and the significance of divergent angles on the behavior of shock wave train are studied using ANSYS Fluent 16 [23]. The atmospheric air is injected at the entrance of the isolator with 220 mm length and 32 mm height. The hydrogen fuel is injected transversely from the either sides of the wall at a distance of 232.8 mm of the inlet of the computational domain.

2.2 Simulation methodology

The commercial software ANSYS Fluent 16 [23] was used to simplify two-dimensional compressible fluid flow by considering the density-based solver with standard K-ε turbulence model, Reynolds-averaged Navier Stokes equation with finite volume method was considered. The species transport model with single step volumetric reaction mechanism is considered to simplify the combustion model (finite rate/eddy dissipation model) [24–27]. To maintain the proper mixing and optimizing the combustion phenomena in supersonic flow RANS approach is the most effective and faster method. The standard K-ε turbulence model is chosen due to its ability of simplifying the negative pressure gradient in the case of transverse injection flow field.

The appropriate governing Eqs. (1)–(5) describing the continuity equation, Navier Stokes equation and combustion model for fluid flow is written as [21, 22, 28, 29].

Continuity equation:

$$\frac{\partial \rho}{\partial t} + \frac{\partial (\rho u_j)}{\partial x_i} = 0 \tag{1}$$

Conservation of momentum (Navier–Stokes equation)

$$\frac{\partial(\rho u_i)}{\partial t} + \frac{\partial(\rho u_i u_j)}{\partial x_j} = -\frac{\partial p}{\partial x_j} + \frac{\partial}{\partial x_j}\left[\mu_{eff}\left(\frac{\partial u_i}{\partial x_j} + \frac{\partial u_j}{\partial x_i}\right)\right] + S_{ui} \tag{2}$$

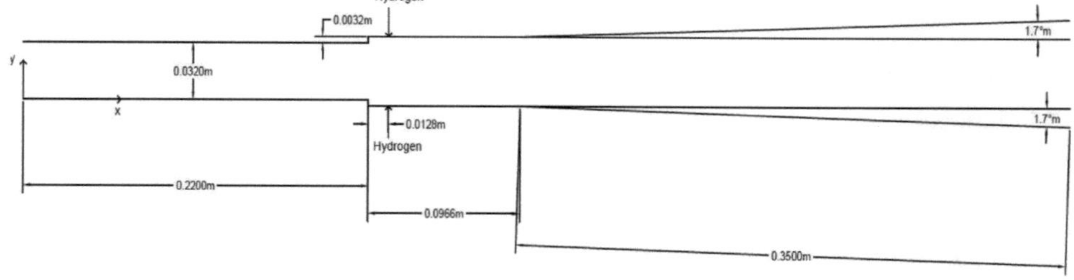

Figure 1.
Schematic diagram of the scramjet combustor.

where the source term S_{ui} includes Coriolis and centrifugal forces

$$S_{ui} = -2\Omega \times U - \Omega \times (\Omega \times r)$$

Conservation of energy equation:

$$\frac{\partial(\rho H)}{\partial t} - \frac{\partial \rho}{\partial t} + \frac{\partial(\rho u_i H)}{\partial x_j} = \frac{\partial}{\partial x_j}\left(k\frac{\partial T}{\partial x_j} + \frac{\mu_i}{\mathrm{Pr}_i}\frac{\partial h}{\partial x_j}\right) + S_E \tag{3}$$

Turbulence transport equations:
K-ε turbulence model and turbulence eddy dissipation equation

$$\frac{\partial(\rho k)}{\partial t} + \frac{\partial(\rho k u_j)}{\partial x_j} = \frac{\partial}{\partial x_j}\left(\left(\mu + \frac{\mu_t}{\sigma_{k3}}\right)\frac{\partial k}{\partial x_j}\right) + \tau_{ij}\frac{\partial u_i}{\partial x_j} - \beta^* \rho k w \tag{4}$$

$$\frac{\partial \rho \varepsilon}{\partial t} + \frac{\partial(\rho u_j \varepsilon)}{\partial x_j} = \frac{\partial}{\partial x_j}\left(\Gamma_k \frac{\partial \varepsilon}{\partial x_j}\right) + \frac{\varepsilon}{k}(C_{\varepsilon 1}P_k - \rho C_{\varepsilon 2}\varepsilon) \tag{5}$$

2.3 Combustion modeling

To simulate the combustion flow dynamics more attention is required as rapid turbulence creation and chemical reaction is required. The species transport model with single step volumetric reaction mechanism is considered to simplify the combustion model (finite rate/eddy dissipation model) which is mainly used in the present research work. The global one step chemical reaction of hydrogen combustion has been considered in this paper for its capability of predicting the overall performance parameters with considerably less computational cost for the scramjet combustor. In this global one step reaction mechanism the rate constants like pre-exponential factor (A) and activation temperature are considered as 9.87×10^8. The one step volumetric reaction mechanism is defined as shown in the Eq. (6):

$$2H_2 + O_2 = H_2O \tag{6}$$

2.4 Boundary condition

An atmospheric air is injected at a velocity of 1200 m/s with a hydraulic diameter of 0.05 m and turbulence intensity of 5%. Hydrogen fuel is injected transversely through the either sides of the walls at x = 0.220 m at a velocity of 900 m/s with a turbulence intensity of 5%. No slip condition and constant heat flux is chosen along the solid surface with standard wall function. Interior combustor zone was chosen for the fluid domain (**Table 1**).

Parameters	Hydrogen Jet	Free-stream jet
Mach number [M]	1.0	4.5
Temperature [K]	1000	1300
Pressure [Pa]	506,625	101,325
C_{H2}	1.0	0
C_{O2}	0	0.21
C_{H2O}	0	0.032

Table 1.
Inlet conditions for hydrogen and air jet.

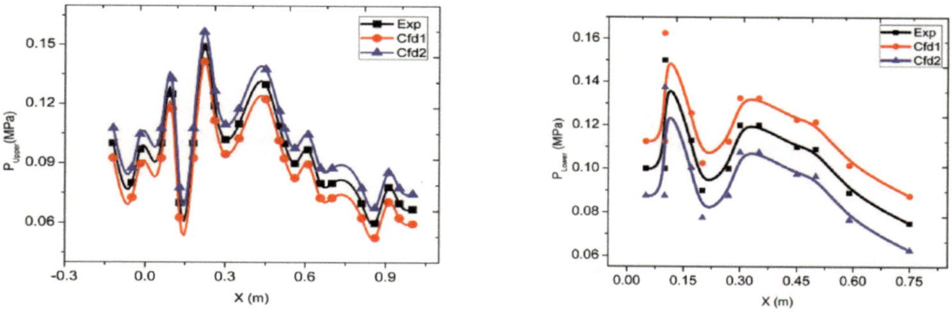

Figure 2.
Wall pressure distribution at mid-plane and bottom wall of the domain.

2.5 Validation of numerical methods and grid independency

To validate the accuracy of numerical results it is required to compare with the experimental results in order to validate the reliability of computational tool. Computational results predominantly depend on the quality of the mesh and size; therefore, it is also necessary to find out the ideal grid size. Initially, computational analysis is carried out to validate the commercial code and simultaneously find out the ideal grid size by considering 634,846 (CFD1, fine mesh) and 384,592 (CFD2, coarse mesh) [25]. The obtained simulated results (**Figure 2**) are then validated by with the experimental data [25] available in open literature and found to be in good qualitative agreement. It is observed that simulation which own fine grid and coarse grid has a good accuracy that the relative error is below 5%. Therefore, the CFD tool can be applied to capture the shock wave reasonably well in terms of both location and strength of the shock wave system [25].

3. Results and discussions

In the present study, the significance of the angle of attack on the wall surface of the domain and the effect of the adverse pressure gradient on the movement of the shock wave train are analyzed. The present analysis focused on different divergent angles, i.e., 0, 0.5, 1 and 1.5° with a standard pressure gradient of 90 kPa, and also with different negative pressures of 80 and 100 kPa with constant cross-sectional area of the isolator is discussed.

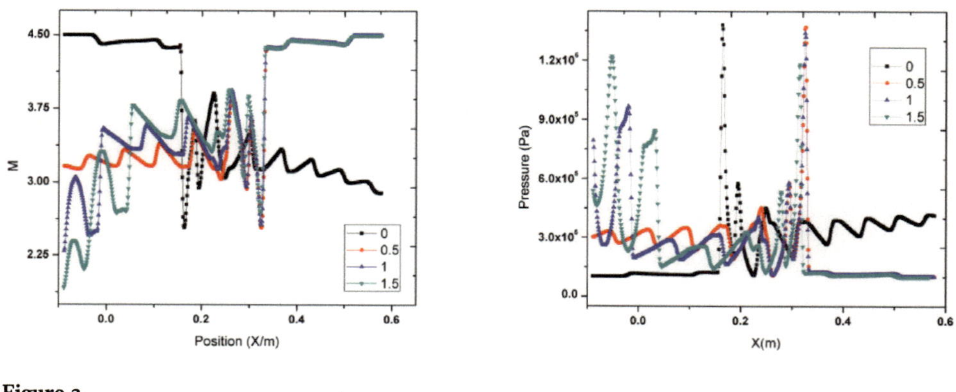

Figure 3.
Mach number and pressure variation along the mid-plane.

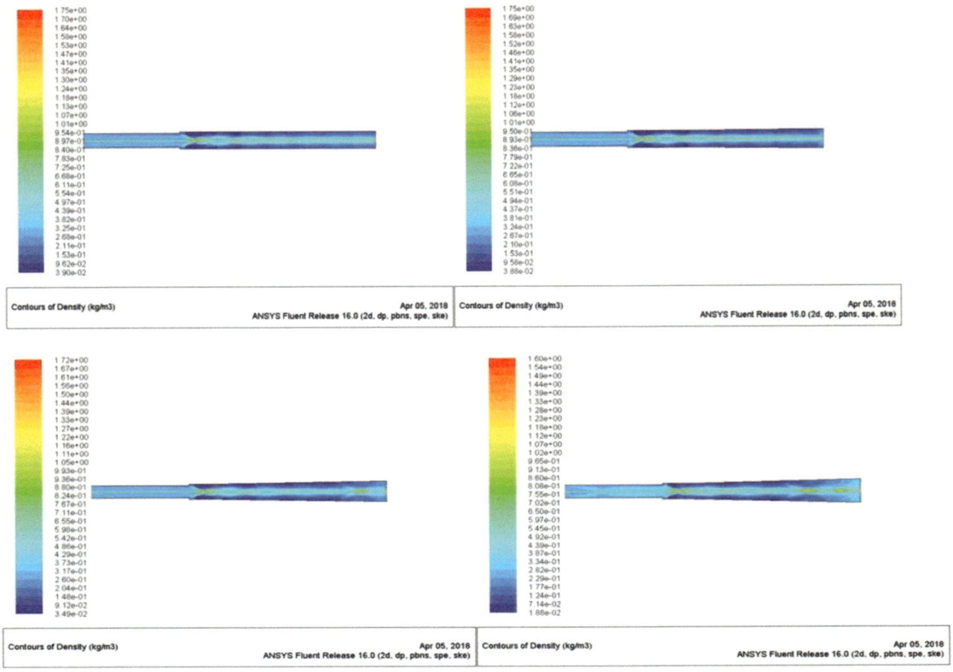

Figure 4.
Density distribution at constant back pressure with variable divergence angles.

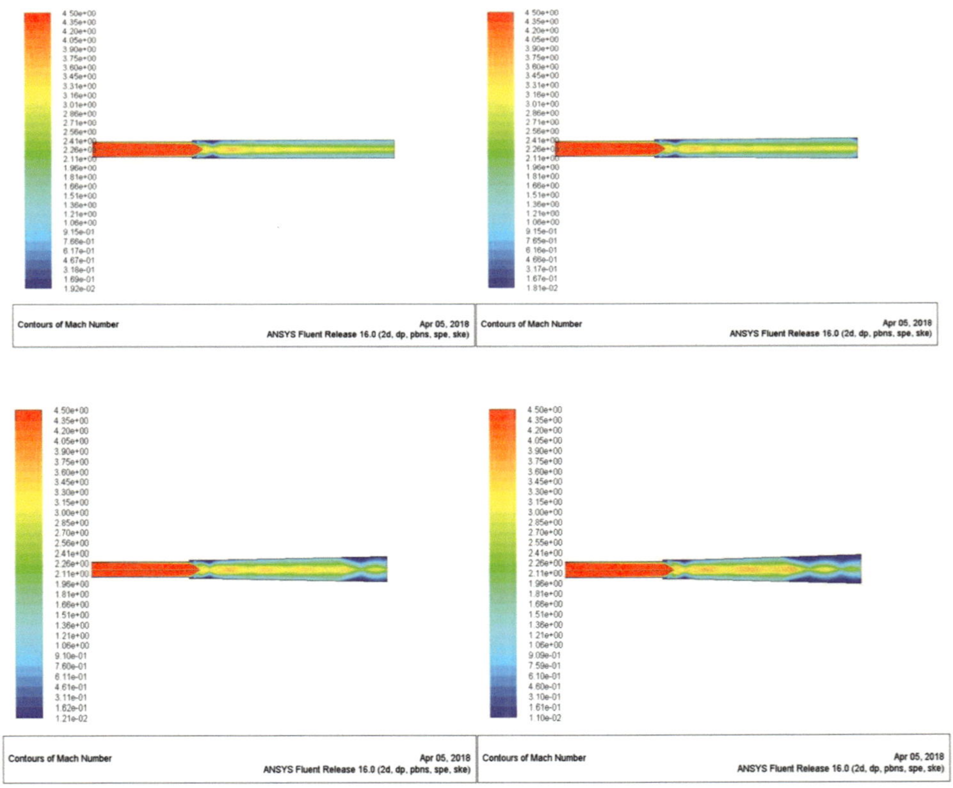

Figure 5.
Mach distribution at constant back pressure with variable divergence angles.

3.1 Impact of divergent angles

In this work, analysis has been carried out at a standard back pressure of 90 kPa with different divergent angles, i.e., 0, 0.5, 1 and 1.5° are considered. The basic purpose of considering the different divergent angles is to analyze the movement of the combustion phenomena inside the isolator. For each divergent angle with a standard adverse pressure gradient of 90 kPa, the contour lines of static temperature, mass density, Mach number and static pressure are measured. From **Figures 3–7**, it is observed that with an increase in the divergent angle the location of the shock train will be moved near to the leading edge of the domain. At a divergent angle of 1.5°, as the flow gets separated the strong expansion wave is generated leading to negative pressure drop at the inception of the Mach wave train. When compared to the divergent angles of 1 and 1.5° in the scramjet isolator, the divergent angle of 0° with constant isolator area supports better back pressure. This is because with an increase in the divergence angle, the shock train generated inside the isolator converts the Mach shock wave into normal shock wave initially and again converts into oblique shock wave. The drawbacks of the normal shock wave generated inside the isolator due to an increase in the divergent angle leads to boundary layer separation on the either sides of the wall of the domain, resulting in the decrease of the intensity of the initial shock wave train. **Figure 3** represents the Mach number and static pressure distribution along the axis of the isolator with different divergent angles. From **Figures 3–7** it is noted that variation in divergence angle leads to stronger shock wave train resulting in rapid pressure losses. The

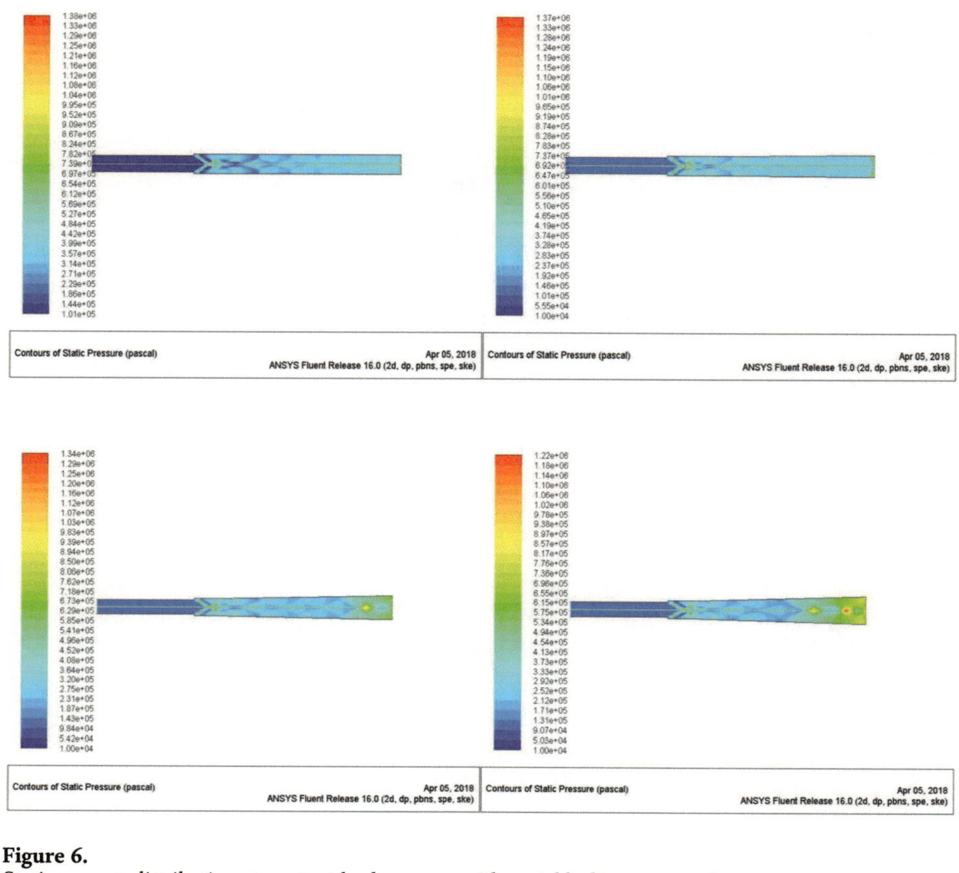

Figure 6.
Static pressure distribution at constant back pressure with variable divergence angles.

impact of the divergent angle on the scramjet combustor has greater significance on the flow field.

3.2 Variation of the back pressures

In the present analysis, the effect of back pressure (i.e., 80, 90 and 100 kPa) on the performance of the domain with uniform cross-sectional area (i.e., divergent angle of 1.5°) are studied. The back pressure of 80 and 100 kPa are identified based on the experimental results by Sun et al. **Figures 8–11** represent the contour lines of static pressure, density, Mach number and temperature along the Mid-plane with

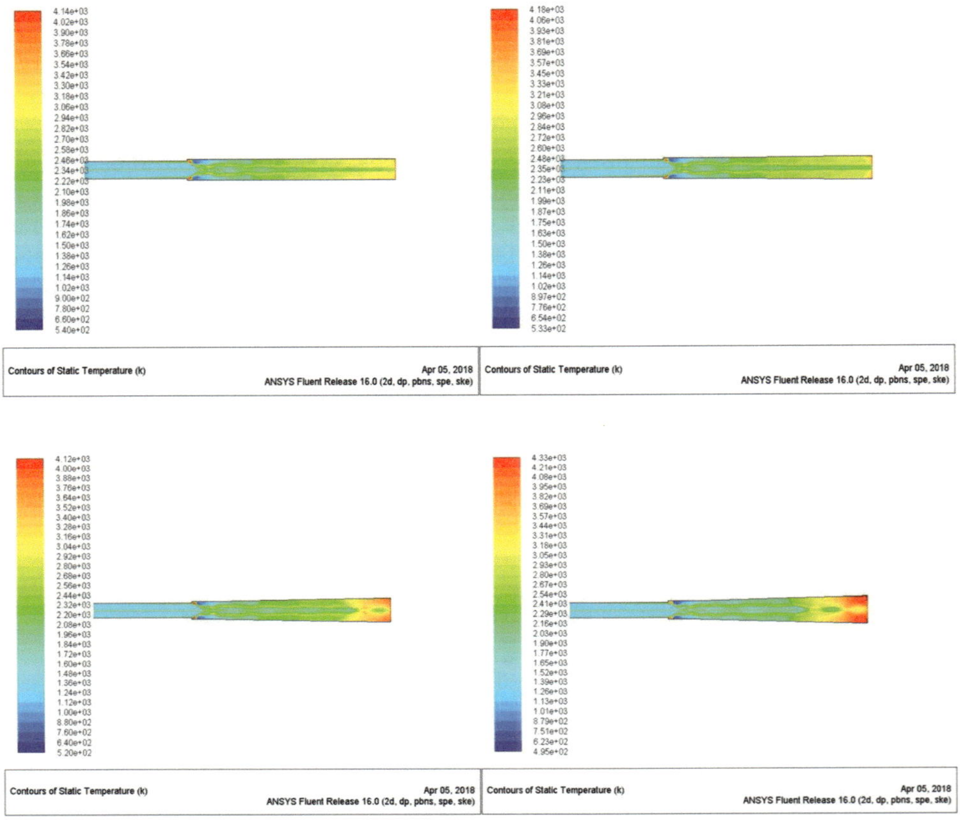

Figure 7.
Static pressure distribution at constant back pressure with variable divergence angles.

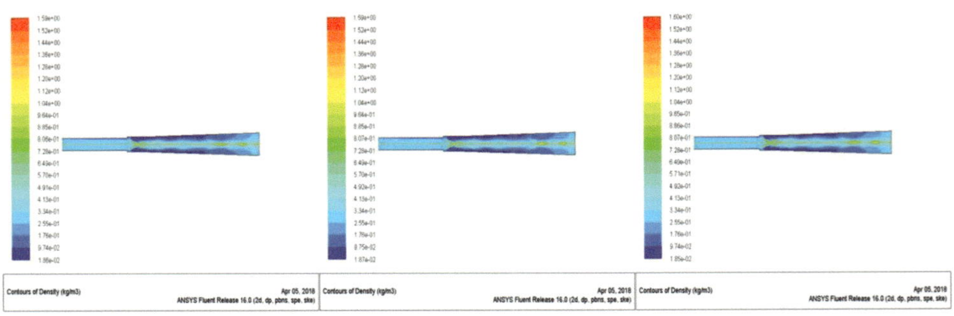

Figure 8.
Density distribution with uniform cross-sectional area with different back pressure.

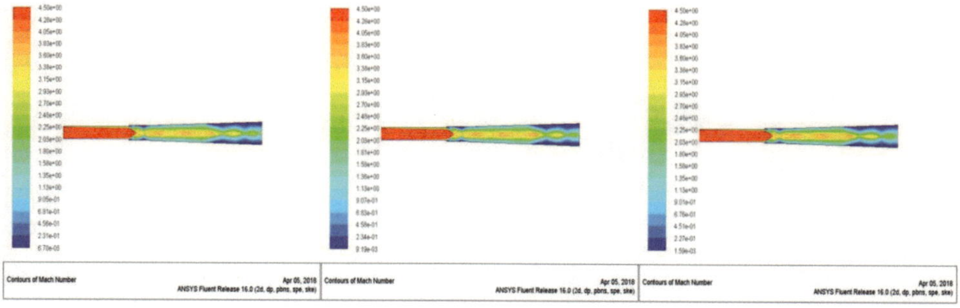

Figure 9.
Mach number distribution with uniform cross-sectional area with different back pressure.

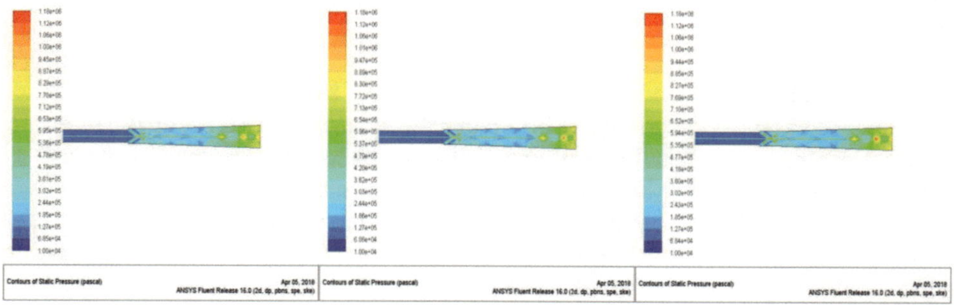

Figure 10.
Static pressure distribution with uniform cross-sectional area with different back pressure.

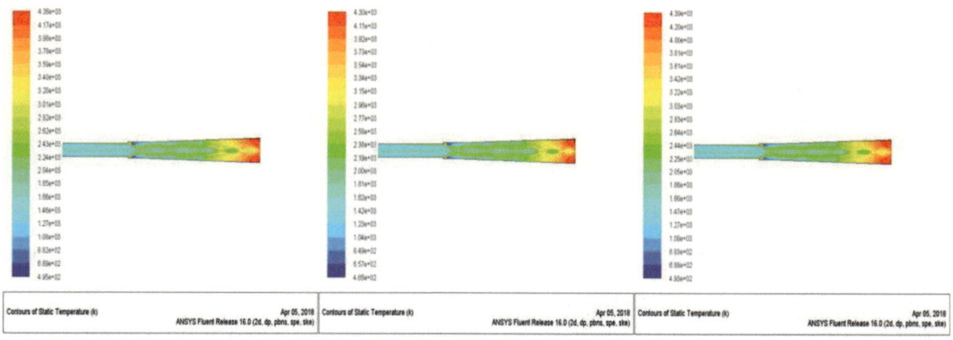

Figure 11.
Static temperature distribution with uniform cross-sectional area with different back pressure.

different back pressure at a uniform cross-sectional area. From the results, it is noted that with an increase in the adverse pressure due to the rapid mixing of the fluid particles, the shock wave train developed lead to intense combustion, the Mach train developed should expand along the length and also moves towards the leading edge of the isolator leading to rapid rise in the pressure along the axis of the isolator so that the pressure at the inlet of the isolator can match the enhanced negative pressure. Additionally, as the shock wave train approaches the supersonic inlet, the unstart conditions are observed at the entrance of the isolator. A strong separation region occurs because of the interaction between shock wave and boundary layer. Due to the increase in the pressure gradient four Mach disks are observed in the domain. With different back pressure of 80, 90 and 100 kPa, symmetric planes of the scramjet isolator the contours of the static temperature,

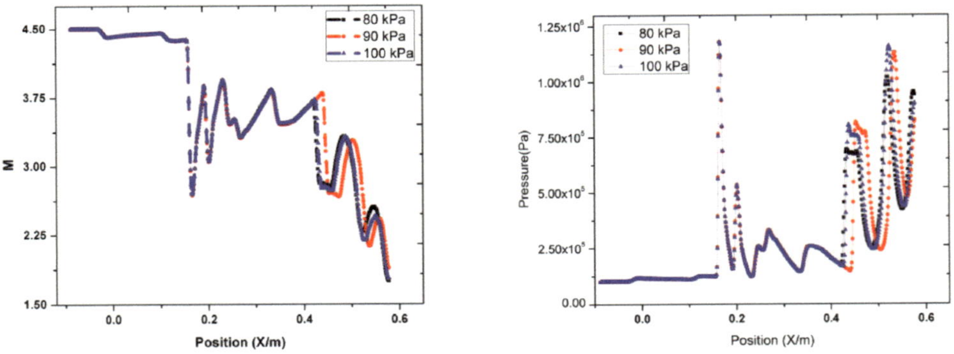

Figure 12.
Mach number and pressure variation along the mid-plane with different back pressure at a uniform cross-sectional area.

pressure, Mach number and density are shown in the figure. From the results it is observed that at a back pressure of 80 kPa in the scramjet isolator, enhancement in pressure is affected due to the presence of the normal or oblique shock waves. Whereas in the case of 90 kPa back pressure, pressure rise is noted far away from the shock wave train due to the intermixing of the disorganized streamlines developed by the shock wave train. Additionally, when compared to downstream mixing region additional enhancement in the static pressure is observed due to the development of upstream shock wave train. For an adverse pressure of 100 kPa, the pressure drops in a given length due to fanno flow is much higher when compared to the increase in pressure due to intermixing and also the peak pressure is observed followed by the sequential reduction in pressure.

From **Figure 12**, it is observed that the static pressure along the axis of the isolator is observed to increase substantially. Also, with different back pressure investigated it is observed that the static pressure distribution does not get influenced significantly because of the Mach wave train in the shock wave region of the isolators.

4. Discussions

The present analysis is focused on the significance of diverging angles and the effect of adverse pressure gradient on the behavior of the shock wave train is analyzed using the simulation. An identical flow inlet is considered at the inception point of the scramjet isolator for different diverging angles, i.e., 0, 0.5, 1 and 1.5° followed by different back pressure, i.e., 80, 90 and 100 kPa have been investigated. The following has been observed. It is observed that the shock wave train has moved near to the leading edge of the isolator with an increase in the divergent angle. At a divergent angle of 1.5°, as the flow gets separated the strong expansion wave is generated leading to the negative pressure drop at the inception of the shock wave train. When compared to the divergent angles of 1 and 1.5° in the scramjet isolator, the divergent angle of 0° with constant isolator area supports better back pressure. With an increase in adverse pressure gradient, because of the intense turbulent combustion, the shock wave train developed should expand along the length and also moves towards the leading edge of the isolator leading to rapid rise in the pressure so that the pressure at the inlet of the domain can match the enhanced negative pressure. A strong separation region occurs because of the interaction between shock wave and boundary layer.

Author details

Santhosh Kumar Gugulothu[1]*, B. Bhaskar[2] and V.V. Phani Babu[3]

1 Department of Mechanical Engineering, National Institute of Technology, Andhra Pradesh, India

2 Department of Mechanical Engineering, GITAM School of Technology, Hyderabad, India

3 Department of Mechanical Engineering, MLR Institute of Technology, Hyderabad, India

*Address all correspondence to: santoshgk1988@gmail.com

References

[1] Xu K, Chang J, Li N, Zhou W, Daren Y. Experimental investigation of mechanism and limits for shock train rapid forward movement. Experimental Thermal and Fluid Science. 2018;**98**: 336-345

[2] Xu K, Chang J, Zhou W, Daren Y. Mechanism of shock train rapid motion induced by variation of attack angle. Acta Astronautica. 2017;**140**:18-26

[3] Chen Z, Yi SH, Ahmed NA, Kong XP, Quan PC. Transient behavior of shock train with and without controlling. Experimental Thermal and Fluid Science. 2015;**66**:79-96

[4] Huang W, Wang Z-G, Pourkashanian M, Ma L, Ingham DB, Luo S-B, et al. Numerical investigation on the shock wave transition in a three-dimensional scramjet isolator. Acta Astronautica. 2011;**68**(11–12):1669-1675

[5] Reinartz BU, Herrmann CD, Ballmann J, Koschel WW. Aerodynamic performance analysis of a hypersonic inlet isolator using computation and experiment. Journal of Propulsion and Power. 2003;**19**(5):868-875

[6] Balu G, Panneerselvam S, Rathakrishnan E. Computational studies on the performance of an isolator for a dual mode scramjet engine. International Journal of Turbo and Jet Engines. 2005;**22**(4):255-264

[7] Fischer C, Olivier H. Experimental investigation of the shock train in an isolator of a scramjet inlet. In: 17th AIAA International Space Planes and Hypersonic Systems and Technologies Conference; 2011. p. 2220

[8] Le DB, Goyne CP, Krauss RH, McDaniel JC. Experimental study of a dual-mode scramjet isolator. Journal of Propulsion and Power. 2008;**24**(5): 1050-1057

[9] Xu K, Chang J, Li N, Zhou W, Daren Y. Preliminary investigation of limits of shock train jumps in a hypersonic inlet-isolator. European Journal of Mechanics - B/Fluids. 2018; **72**:664-675

[10] Tan HJ, Sun S, Huang HX. Behavior of shock trains in a hypersonic inlet/ isolator model with complex background waves. Experiments in Fluids. 2012;**53**(6):1647-1661

[11] Huang H-x, Tan H-j, Sun S, Wang Z-y. Behavior of shock train in curved isolators with complex background waves. AIAA Journal. 2017:329-341

[12] Krishnan L, Sandham ND, Steelant J. Shock-wave/boundary-layer interactions in a model scramjet intake. AIAA Journal. 2009;**47**(7):1680-1691

[13] Carroll BF, Dutton JC. Characteristics of multiple shock wave/ turbulent boundary-layer interactions in rectangular ducts. Journal of Propulsion and Power. 1990;**6**(2):186-193

[14] Wagner JL, Yuceil KB, Valdivia A, Clemens NT, Dolling DS. Experimental investigation of unstart in an inlet/ isolator model in Mach 5 flow. AIAA Journal. 2009;**47**(6):1528-1542

[15] Xu K, Chang J, Zhou W, Yu D. Mechanism and prediction for occurrence of shock train sharp forward movement. In: 51st AIAA/SAE/ASEE Joint Propulsion Conference; 2015. p. 3747

[16] Li N, Chang J-T, Xu K-J, Yu D-R, Bao W, Song Y-P. Prediction dynamic model of shock train with complex background waves. Physics of Fluids. 2017;**29**(11):116103

[17] Erdem E, Kontis K, Johnstone E, Murray NP, Steelant J. Experiments on

transitional shock wave–boundary layer interactions at Mach 5. Experiments in Fluids. 2013;**54**(10):1598

[18] Geerts J, Yu K. Experimental characterization of isolator shock train propagation. In: 18th AIAA/3AF International Space Planes and Hypersonic Systems and Technologies Conference; 2012. p. 5891

[19] Koo H, Raman V. Large-eddy simulation of a supersonic inlet-isolator. AIAA Journal. 2012;**50**(7):1596-1613

[20] Shang JS, Hankey WL Jr, Law C. Numerical simulation of shock wave-turbulent boundary-layer interaction. AIAA Journal. 1976;**14**(10):1451-1457

[21] Morgan B, Duraisamy K, Lele SK. Large-eddy simulations of a normal shock train in a constant-area isolator. AIAA Journal. 2014;**52**(3):539-558

[22] Versteeg HK, Malalasekera W. An Introduction to Computational Fluid Dynamics: The Finite Volume Method. England: Pearson Education; 2007

[23] Fluent Inc. Guide, FLUENT User's. Version 6.2.16. Lebanon, New Hampshire: Fluent Inc.; 2005

[24] Gugulothu SK. Significance of fluid flow characteristics in the hydrogen-fuelled scramjet combustor using flame stabilisation analysis. Australian Journal of Mechanical Engineering. 2019;**2**:1-10

[25] Gugulothu SK, Nutakki PK. Dynamic fluid flow characteristics in the hydrogen-fuelled scramjet combustor with transverse fuel injection. Case Studies in Thermal Engineering. 2019;**14**:100448

[26] Gugulothu SK. A systematic literature review based on different fuel injection strategies used in scramjet combustors. Heat Transfer - Asian Research

[27] Dharavath M, Manna P, Sinha PK, Chakraborty D. Numerical analysis of a kerosene-fueled scramjet combustor. Journal of Thermal Science and Engineering Applications. 2016;**8**(1):011003

[28] Huang W, Liu W-d, Li S-b, Xia Z-x, Liu J, Wang Z-g. Influences of the turbulence model and the slot width on the transverse slot injection flow field in supersonic flows. Acta Astronautica. 2012;**73**:1-9

[29] Soni RK, Ashoke D. Investigation of strut-ramp injector in a Scramjet combustor: Effect of strut geometry, fuel and jet diameter on mixing characteristics. Journal of Mechanical Science and Technology. 2017;**31**(3):1169-1179

Chapter 2

A New Combustion Method in a Burner with Three Separate Jets

Mohamed Ali Mergheni, Mohamed Mahdi Belhajbrahim, Toufik Boushaki and Jean-Charles Sautet

Abstract

Oxy-flames from burners with separated jets present attractive perspectives because the separation of reactants generates a better thermal efficiency and reduction of pollutant emissions. The principal idea is to confine the fuel jet by oxygen jets to favor the mixing in order to improve the flame stability. This chapter concerns the effect of equivalence ratio on characteristics of a non-premixed oxy-methane flame from a burner with separated jets. The burner of 25 kW power is composed with three aligned jets, one central methane jet surrounded by two oxygen jets. The numerical simulation is carried out using Reynolds Average Navier-Stokes (RANS) technique with k-ε as a turbulence closure model. The eddy dissipation model is applied to take into account the turbulence-reaction interactions. The study is performed with different global equivalence ratios (0.7, 0.8 and 1). The validation of the numerical tools is done by comparison with experimental data of the stoichiometric regime ($\Phi = 1$). The two lean regimes of $\Phi = 0.7$ and 0.8 are investigated only by calculations. The velocity fields with different equivalence ratio are presented. It yields to increase of longitudinal and transverse velocity, promotes the fluctuation in interaction zone between fuel and oxygen also a better mixing quality and a decrease of the size of the recirculation zone.

Keywords: oxy-flame, turbulent, separated-jets, equivalence ratio, flame stability

1. Introduction

This chapter concerns the numerical simulation and the PIV measurements on oxy-fuel burners with three separated jets. The mixing, dynamic and the temperature fields for both reacting and non-reacting flows are investigated.

Industrial systems with combustion phenomena such as burners, aeronautical engines and gas turbines are subject to increasingly important constraints, both economically (cost reduction, improved performance, etc.) than on the environmental level (reduction of pollutant emissions), requiring the development of new techniques to respond effectively to these industrial constraints. The development passes by new method of combustion in order to reduce pollutant emissions and fuel consumption, as well as by the improvement of flame stability [1, 2]. In previous studies, significant reductions of nitric oxide emissions have been successfully achieved by using low NOx technologies or oxy-combustion systems [3].

In air combustion, nitrogen leads to high fuel consumption and low combustion efficiency because nitrogen in the air acts as energy ballast. The substitution of air with pure oxygen leads to an increase in the laminar combustion rate up to 1300%, improves the thermal efficiency, increases the adiabatic flame temperature (2200 K for CH4-Air, 3090 K in oxy-combustion) reduce fuel consumption by 50% and, from an environmental point of view, reduce the formation of nitrogen oxides by up to 95% [4].

The flames from multiple jets aligned have used in many industrial installation. Several studies have been published on the dynamic properties of non-reacting multiple jets [5–9]. Lee et al. [10] have studied the geometry parameters of diffusion flames and giving a number of variables such as the number of jets and the distance between the jets. Lenze et al. [11] have studied the influence of three and five non-premixed flames, with town gas and natural gas burners. Their measurements concern flame width, flame length and concentrations in confined and free multiple flames.

A new generation of highly separated fuel and oxidant injection burners is of great interest to industrialists. The idea of this burner consists of separating combustible and oxidant to dilute the reactants with combustion products before the mixing of the reactants [12–14].

For this new combustion in a burner with three separated jets, the separation of jets provides a high dilution of reactants by combustion products in the combustion chamber. Consequently, this dilution decreases the flame temperature and decline in NOx production. In the literature, it has been proven that the separation of reactants are capable to change the flow structure, the flame characteristics, generates a better thermal efficiency and as well as reduction of pollutant emissions [15–18].

Salentey et al. [16] was interested in the characterization the flames from multiple jets aligned through dynamic properties (speed of the jets and distance injectors) and the flame topology (stability, length, blow ...). Lesieur et al. [14] has studied numerically the characteristics of a burner with three jets, focusing on the mixing of the jets, their dynamics and the pollutant emissions. Boushaki et al. [12] was interested on two main areas for flow, passive control with changing the diameter of the burner in order to affecting the dynamics flow; and active control requiring external energy intake through actuators while retaining the geometry of the combustion chamber.

The present chapter reports the results of a numerical and experimental investigation of the dynamic field on a burner with 25 kW power composed of three jets, one central jet of natural gas and two side jets of pure oxygen [19, 20]. One control systems, passive, is added to the basic burner to ameliore the combustion process to ensure the stabilization of flame and as well as pollutant reductions. The passive control is based on the inclined of side oxygen jets towards the central natural gas jet in burner with three separated jets.

Few works, are investigated the effect of equivalence ratios (in lean regime) on characteristics of non-premixed oxy-methane flames from burner with separated jets. However, the aim of this contribution is to investigate numerically the effect of different equivalence ratio on the combustion characteristics of a diffusion methane oxy-flame in a stabilized separated burner.

The mixture of hydrogen and natural gas is a new mixed fuel. The use of a mixed mixture of fuel and hydrogen has the advantage of modifying very effectively the properties of the fuel while preserving the distribution facility. Due to this, the high molecular diffusivity of hydrogen, the extended flammability limits, the high laminar flame speed and the low ignition energy, the addition of hydrogen in the fuel makes it possible to work in a combustion poor. Increasing flammability limits in the presence of hydrogen offset the adverse effects of poor combustion such as local extinctions, radiation energy losses, and flame stretching [21].

2. Burner configuration and experimental set-up

2.1 Basic configuration of the burner

The configuration of the burner illustrated in **Figure 1** consists in separating the fuel and oxygen per injection in order to increase the dilution of the reactants with the combustion products before the mixing of the reagents.

This burner consists of three non-ventilated jets, one central with internal diameter dg equal 6 mm that contains the fuel and two side jets with internal diameter dox equal 6 mm contain pure oxygen. Boushaki et al. [22] have studied this three-jet burner configuration. The separation distance between the jets (S) used is 12 mm. The gas density equal to 0.83 kg m^{-3} and the oxygen is supplied by liquid air with a purity of 99.5% with a density of 1.354 kg m^{-3} (at 1 atm and at 15° C). The thermal power (P) of the burner is equal to 25 kW, therefore the flow rate and the output speed of the natural gas are respectively m_{ng} = 0.556 g s^{-1} and U_{ng} = 27.1 ms $^{-1}$.

The first study in this document is the control technique, consists in inclining the side oxygen jets towards the natural gas jet as shown in **Figure 1**. The angle of oxygen jets (Θ) compared to the vertical direction varies from 0 to 30° (0, 10, 20, and 30°), however, we will present the effect of angle of the side oxygen jets on the dynamic fluid, for many detail you can see [12].

The combustion is carried out inside a square chamber of 60 × 60 cm^2 section and a height of 1 m. The side walls are water cooled and refractory lined inside the combustion chamber. a converging 20 cm high and a final section of 12 × 12 cm is placed at the end of the chamber to limit the entry of air from above. In order to allow optical access to all flame zones, six windows are provided in each face of the chamber.

The Particle Image Velocimetry (PIV) was used as a measurement technique to characterize the experimental dynamic field. The PIV technique requires a laser sheet that clarifies the flow area studied a CCD camera, control equipment and an acquisition PC. The laser used is the Nd-YAG Bi-pulse with frequency 10 Hz and wavelength of 532 nm. The laser chain used is composed by a first divergent cylindrical lens and then by a second convergent spherical lens. The Mie signal emitted by the particles is collected by CCD camera of type a Lavision FlowMaster (12-bit dynamic and resolution 1280 × 1024 pixels).

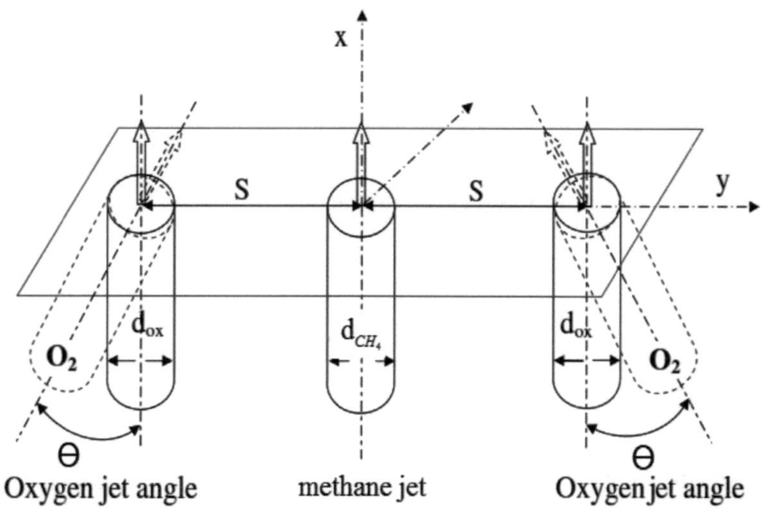

Figure 1.
Schematic view of the burner.

3. Numerical method

The steady equations for conservation of mass, momentum, energy and species have been used in this numerical simulation. The second order equations for turbulence kinetic energy κ and its rate of dissipation ε have been used to modulate the turbulence. The general form of the elliptic differential equations for an axisymmetric flow is given by Eq. (1).

Here S_Φ is the source term and Γ_Φ is the transport coefficient.

$$\frac{\partial}{\partial x}(\rho U\Phi) + \frac{1}{r}\frac{\partial}{\partial r}(r\rho V\Phi) = \frac{\partial}{\partial x}\left(\Gamma_\Phi \frac{\partial\Phi}{\partial x}\right) + \frac{1}{r}\frac{\partial}{\partial r}\left(r\Gamma_\Phi \frac{\partial\Phi}{\partial r}\right) + S_\Phi \qquad (1)$$

where ρ is the density, P is the mean pressure and μ is the viscosity.

μ_e is the effective viscosity is determined from $\mu_e = \mu + \mu_t$, where $\mu_t$is the turbulent viscosity, which is derived from the turbulence model and expressed by: $\mu_t = C_\mu \rho \frac{\kappa^2}{\varepsilon}$.

The Finite Eddy Dissipation Model (EDM) is used to simulate the turbulence/chemistry interaction. This model is based on the hypothesis that the chemical reaction is fast in relation to the transport processes of the flow.

Table 1 summarizes the volumetric flow rates, the velocities, Reynolds number respectively of methane and oxygen of equivalence ratio (0.7, 0.8 and 1).

Reynolds Number is defined by the following equation:

$$Re = (\rho\, d_{gn}\, U)/\mu \qquad (2)$$

A global equivalence ratio can be defined as the molar ratio of methane and oxidant at the injection to molar ratio methane and oxidant in stoichiometric conditions, as:

$$\phi = h\left(\frac{Q}{Q_{O2}}\right) \Big/ \left(\frac{Q_{CH4}}{Q_{O2}}\right)_{stoichio} \qquad (3)$$

where Q is the volumetric flow rate.

The second numerical study in this document is the effect of hydrogen to dynamic of flame. One of the jet transports the fuel, natural gas + hydrogen, and the other the pure oxygen. The values of the flow rates of fuel and the exit velocities are regrouped in **Table 2**.

We shall consider an overall irreversible reaction between methane/hydrogen and pure oxygen:

$$(1-\alpha)\,\Phi CH_4 + \alpha\Phi H_2 + a\Phi O_2 \rightarrow b\Phi CO_2 + c\Phi H_2O \qquad (4)$$

α is the percentage of hydrogen and written as:

$$\alpha = \frac{\%H_2}{\%H_2 + \%CH_4} \qquad (5)$$

Configuration	ϕ	$\dot{Q}_{CH_4}\left[\frac{l}{s}\right]$	$\dot{Q}_{O_2}\left[\frac{l}{s}\right]$	$U_{ng}\left[\frac{m}{s}\right]$	$U_{O_2}\left[\frac{m}{s}\right]$	Re_{gn}
Confi 1	1	0.767	1.534	27.13	27.13	12,272
Confi 2	0.8	0.767	1.917	27.13	33.86	12,272
Confi 3	0.7	0.767	2.19	27.13	38.69	12,272

Table 1.
Dynamic conditions of the burner.

	Φ =1				
	P=25 kW				
	$\dot{m}_{CH4}$ (g.s^{-1})	$\dot{m}_{H2}$ (g.s^{-1})	$\dot{m}_{O2}$ (g.s^{-1})	U_{ng} (m.s^{-1})	U_{O2} (m.s^{-1})
0% H_2	0.49	0	2.07	27.07	27.06
20% H_2	0.46	0.012	2.03	31.3	26.66
40% H_2	0.40	0.02	1.98	37.07	25.9

Table 2.
Flow rates and exit velocities of fuels and oxygen.

Table 2 summarizes the parameters of this numerical study including methane, hydrogen and oxygen flow rates, velocities, percentage of hydrogen and equivalence ratio.

Fluent 6.3.2 is used to solve the steady equations for conservation of mass, momentum, energy and species. The finite volume method is used with second order upwind. In fact, convergence criterion of residuals for energy equation and for all other equations equal respectively 10^{-6} and 10^{-3}. The GAMBIT is used to construct the grid; the computational domain has been extended 100 cm in the axial direction and 30 cm in the radial direction. A total number of 28,700 quadrilateral cells were generated using non-uniform grid spacing to provide an adequate resolution near the jet axis and close to the burner where gradients were large.

The axial velocity profile at the inlet, of the methane is supposed constant. At the axis of symmetry, $r = 0$, $V = 0$ and $\partial\Phi/\partial r = 0$ $(\Phi = U, \kappa, \varepsilon)$. At the outlet, the fully-developed condition of pipe flow is adopted $\partial\Phi/\partial x = 0$ $(\Phi = \mathrm{U}, \mathrm{V}, \kappa, \varepsilon)$. The velocities are assumed to be zero at the wall, and these no-slip boundary conditions are appropriate for the gas. These equations, called "wall functions," are introduced and used in finite difference calculations at near-wall points.

4. Inclined effects on dynamic

4.1 PIV measurements on burners with inclined jets

The mean velocity fields carried out by PIV in non-reacting flow are represented on **Figure 2**. From initial state where Θ = 0° to inclined state where Θ = 30°, the dynamic field changes with the change of flow structure. The jets fusion point

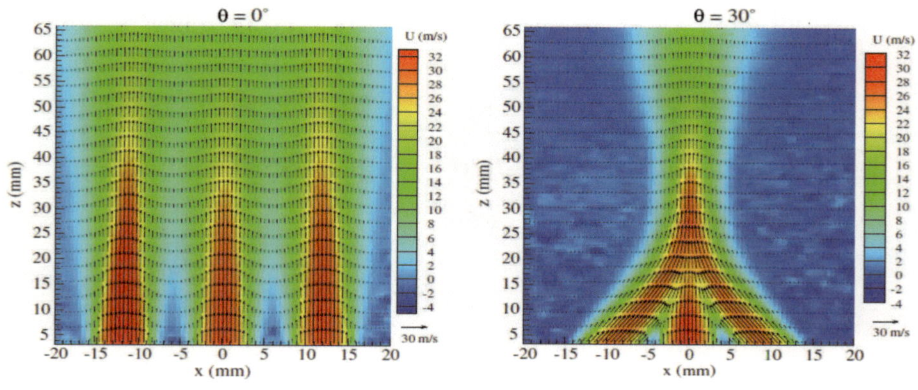

Figure 2.
Mean velocity fields for jet oxygen angle 0° and 30° (longitudinal velocity in color scale) in non-reacting flow.

becomes closer the burner by increasing of the slope of jets. The interaction of jets starts at about 15 mm for Θ = 0°, at z = 25 mm for Θ = 30°.

4.2 Velocity distribution and current lines

Figure 3 shows the distribution of velocity and the current lines in the combustion chamber near the burner with different inclined jets of oxygen. In the part separating the different jets (dark blue), velocity is negative because of the recirculation of the jets. The existence of two zones of recirculation is observed with different directions of rotation, which explains the appearance of the negative velocity. It is noted on the one hand that the recirculation zone decreases with the increase of Θ from 0 to 20°. This is very remarkable near the jet of oxygen. On the other hand, it can be observed that the recirculation zones appear outside the jet of air (see the lines of currents). The perturbation of velocity distribution increases with the increase of Θ. This perturbation is accompanied by an acceleration of the combination of different jets and consequently a faster combustion reaction, which explains the increase in velocity with the increase of Θ.

4.3 Temperature distribution

The distributions of the temperature in the combustion chamber with different inclined jets are represented in **Figure 4**. It is clear here that the flame exists in the mixing zone of methane and oxygen, which represents the reaction zone. This zone is modified with the variation of the angle Θ. If we assume that the length of the flame is defined by the red color of the flame distribution, we can conclude that the length of the flame decreases with the increase of the angle Θ from 0.44 m for an inclination of 0° to 0.29 m for an inclination of 20° of the jet of oxygen.

Figure 5 represents the evolution of the axial temperature at y = 0 mm with different angles of injection Θ of the oxygen jets. Firstly, it is observed that the

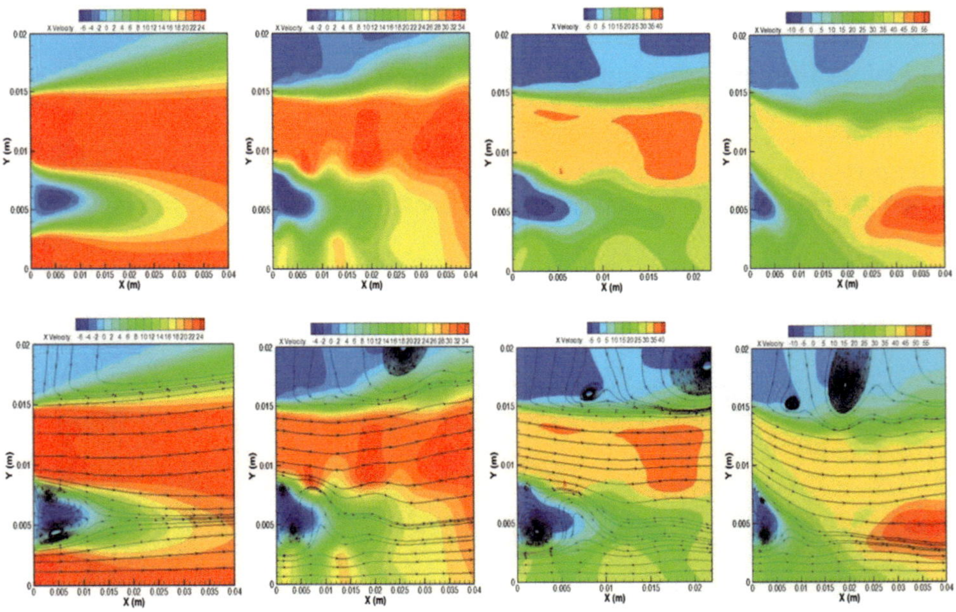

Figure 3.
Velocity distribution and current lines in the combustion chamber.

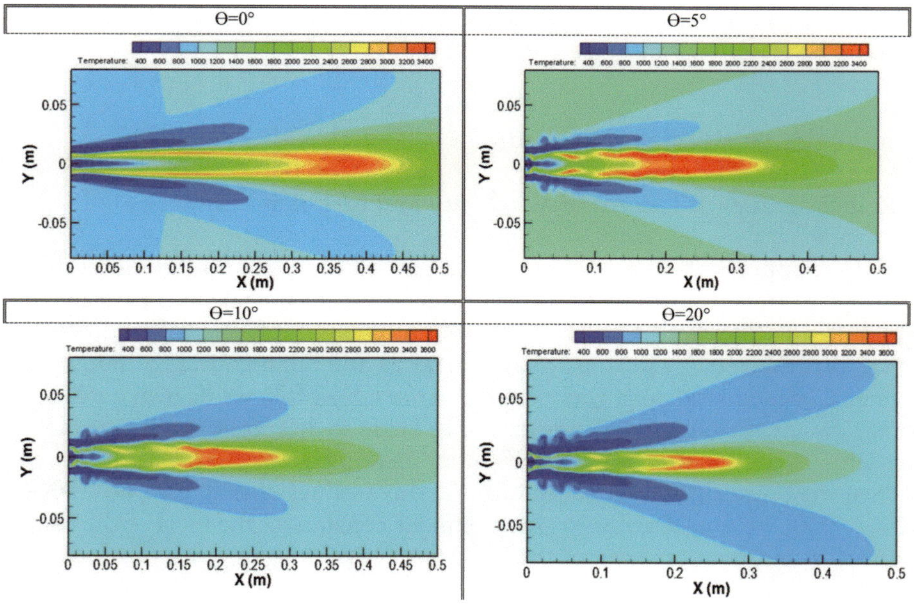

Figure 4.
Temperature distribution in the combustion chamber.

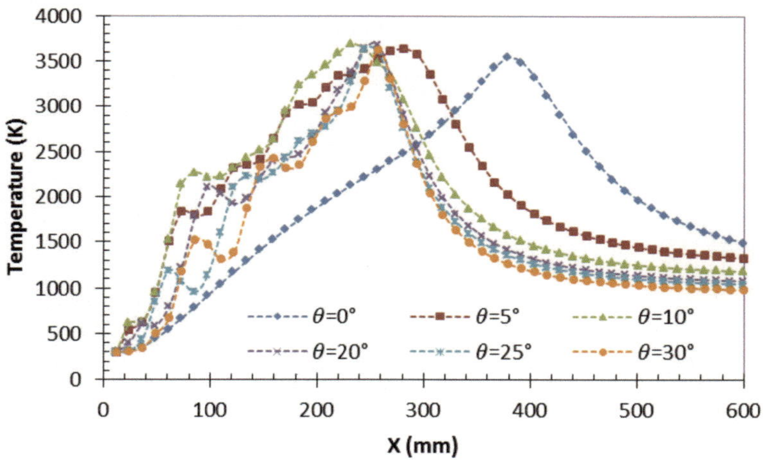

Figure 5.
Axial distribution profiles of temperature at y = 0 mm with variation of the angle Θ.

temperature increases by moving away from the burner to a maximum value and then begins to decrease. For example for Θ = 0° the temperature increases from 300 K near the burner up to 3500 K at a height of 380 mm. This zone of increase presents the mixing zone of the reactants methane/oxygen. The second zone is the reaction zone where the temperature reaches its maximum. The third zone is where the temperature gradually decreases and which presents the plume of the flame. The second interpretation is that the flame reaches its maximum faster while the angle of injection Θ increases. Indeed, the temperature reaches its maximum at a height of 390 mm for Θ = 0° by contrast, it reaches its maximum at 260 mm for Θ = 30°. This interpretation leads to conclude that the length of the flame decreases with the increase in the angle of the oxygen jets. This result is in good agreement with the result of Boushaki [22], which showed that the average flame length decreases when the angle of oxygen jets increases such that its value is about 500 mm for Θ = 0° and decreases until 220 mm for Θ = 30°.

5. Equivalence effects on dynamic

5.1 Radial profiles of longitudinal velocity and turbulence intensity

The radial profiles of the mean longitudinal velocity (U) at different section (x/D = 1.66, x/D = 8.33 and x/D = 16.66) and for three equivalence ratios are represented in **Figure 6**. A classical behavior of the multiple jets is found, one notices that the velocity profile presents maxima and minima corresponding to the three jets. In the initial zone (near the burner) each jet follows its own evolution, further downstream these velocity extremes begin to disappear to form a single maximum located in the middle of the inner mixing layer.

Near the burner (x/D = 1.66) and for the three values of richness (Φ = 1, Φ = 0.8 and Φ = 0.7), we note that the velocity remains constant at the level of the central jet and it increases at the level of the lateral jet (jet of oxygen) with the decrease of the wealth. It should be noted that for Φ = 1 and Φ = 0.7 the mean longitudinal velocities are equal to 27 m/s and 38.57 m/s, mean velocity show an increase of 30%. In the case x/D = 16.66, the velocity profiles are slightly flattened, more open which improves the mixing of the reagents.

The influence of the equivalence ratio on the longitudinal velocity U is significant less. From an aerodynamics point of view, the decrease of equivalence ratio

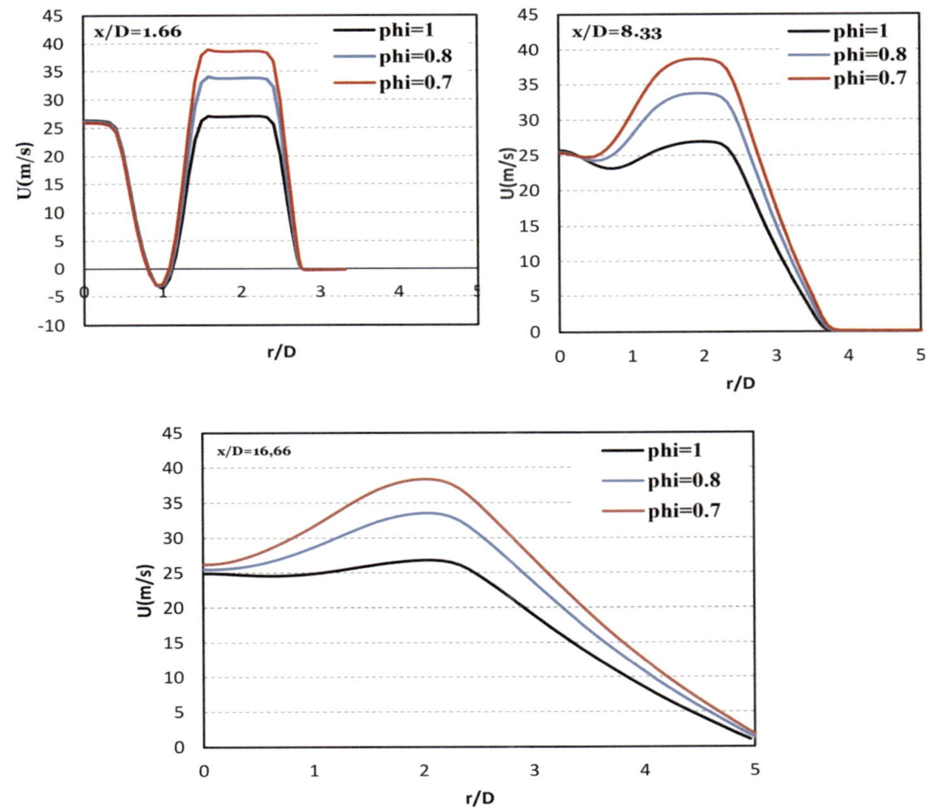

Figure 6.
Radial profiles of longitudinal velocity at different positions from the burner.

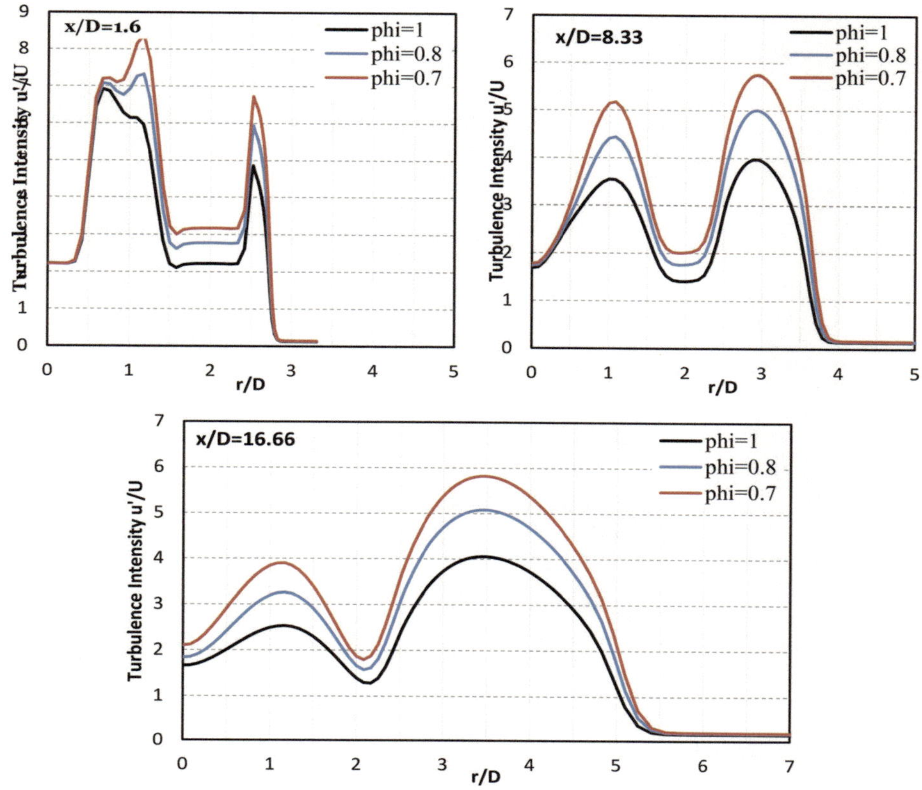

Figure 7.
Turbulence intensity at different positions from the burner.

modifies the longitudinal velocity of flow near to the burner but keep the flow velocity behavior in the combination zone.

Figure 7 shows, the radial profiles of the turbulence intensity, u'/U, on function to the equivalence ratio and on sections different, x/D = 1.66, 8.33 and 16.66. In the case x/D = 1.66 we see two peaks of fluctuations in u' 'of the order of 7 m/s, one at the center corresponding to the mixing layer of the central jet and one from the central jet corresponding to the mixture layer of side jet. These peaks of fluctuations fade along the flow with decreasing longitudinal velocity as the jets interpenetrate.

5.2 Radial profiles of temperature

The distribution of the temperature in sections different inside the combustion chamber, is shown in **Figure 8**. In the near of the burner, the temperature present one peak with maximum value equal to 3000°C. For a richness equal to 1 ($\Phi = 1$) the maximum temperature of the adiabatic flame is T = 3000 K and for a richness 0.7 the maximum temperature equals approximately T = 3300 K. Therefore, the peaks represent the zone of the reaction between the fuel and oxidant after mixing and the region between the peaks represents the area of the fuel which is not yet burned and the reaction takes place at the interface of jets between fuel and oxidant. At x/D = 8.33 and for three equivalence ratio the peak temperature is observed equal to 3200°C and the temperature profile keep constant. Far from the burner when the richness decreases, an increase in the temperature in the flame zone is observed from 3000 to 3500°C which makes it possible to improve the heat transfers and makes it possible to have a better thermal efficiency.

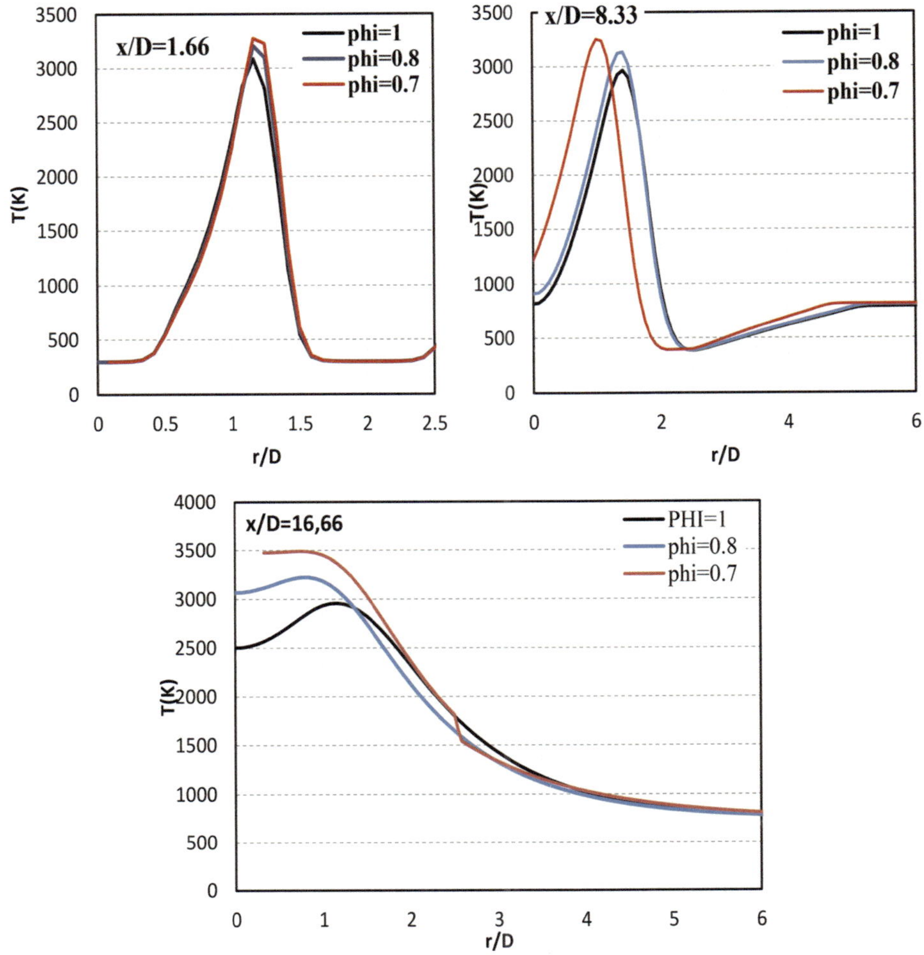

Figure 8.
Radial profiles of temperature at different positions from the burner.

6. Effect of H_2 addition on the dynamic field and temperature

6.1 Longitudinal velocity field

Figure 9 shows the longitudinal velocity field for different percentages of hydrogen. By comparing the four fields, the potential cone for three configurations with the addition of hydrogen percentages ($\alpha_{H_2} = 20\%, 40\%\ and\ 60\%$) are smaller than with pure methane. This indicates that the increase in velocity of the jet when adding hydrogen to the fuel reduces the size of the potential cone allowing the reactants to interact more effectively with oxygen and the ambient fluid. Note also that the addition of hydrogen reduces the height of the mixing area and consequently the flame height due to the high molecular diffusivity and low density of hydrogen promoting interaction between the jets more rapidly and the flow features to the behavior of a single jet.

6.2 Radial profiles of temperature

Figure 10 illustrates the radial temperature profiles for multiple hydrogen percentages in the fuel mixture. This figure shows that the substitution of a fraction of

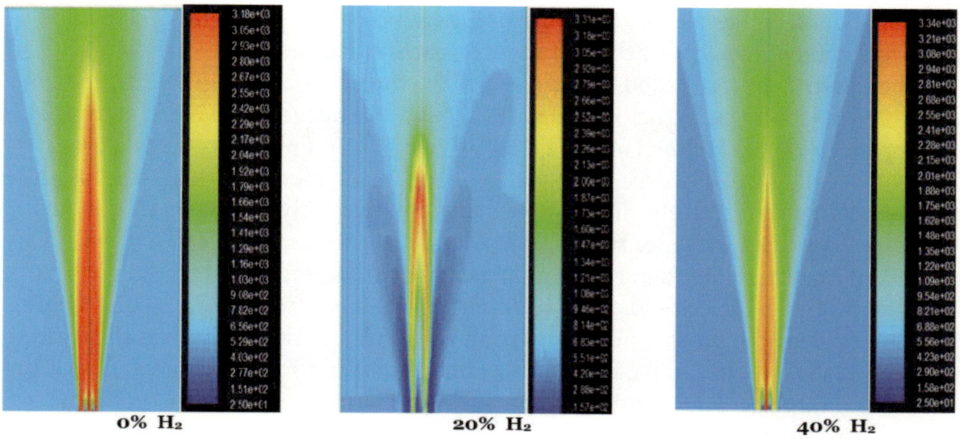

Figure 9.
Longitudinal velocity field for 0% H_2, 20% H_2 and 40% H_2.

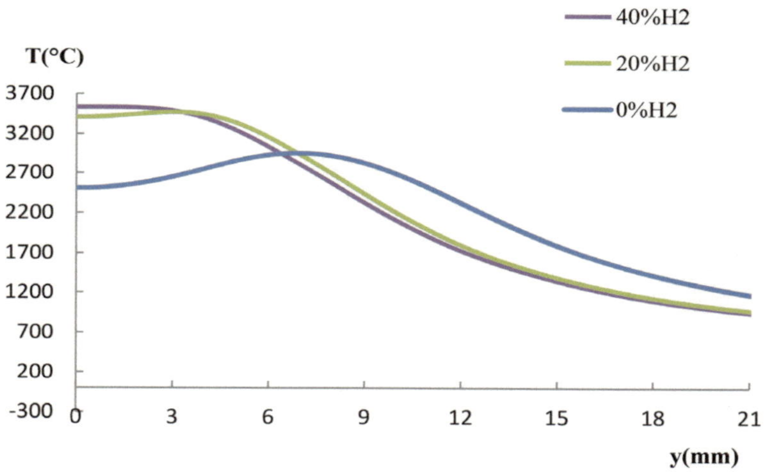

Figure 10.
Radial profiles of the temperature for 0% H_2, 20% H_2, 40% H_2 and 60% H_2.

methane with hydrogen promotes the increase of the flame temperature because the calorific value of this compound is much higher than that of methane. For example, the addition of 20% hydrogen increases the temperature 2500–3400 K (up to 900 K), and the addition of 60% hydrogen increases the temperature until 3700 K (1200 K more than in pure methane combustion).

7. Conclusion

In this chapter, a new combustion technique in a burner with three separated jets is used. The idea of this burner consists of separating combustible and oxidant to dilute the reactants with combustion products before the mixing of the reactants. This type of burner has a great interest for the industry and the sizing of these burners requires a good understanding of the mechanisms controlling the stabilization of the flame, the release of heat and the production of pollutants.

The Particle Image Velocimetry PIV is the technique used in experimental study in non-reacting flow and reacting flow inside the combustion chamber. The Reynolds Average Navier-Stokes (RANS) method is used in this numerical simulation with Realizable k-ε as a turbulence closure model. The eddy dissipation model is applied to take into account the turbulence-reaction interactions.

The passive control added to the basic of the burner is based on the inclined of side oxygen jets towards the central natural gas jet in burner with three separated jets. From Θ = 0° to inclined state Θ = 30°, the jets fusion point becomes closer the burner as well as the dynamic field changes. The result shows that the inclination of the jets affects significantly the flow field and consequently the flame behaviour.

The effect of equivalence ratio and hydrogen on characteristics of a non-premixed oxy-methane flame from a burner with separated jets is studied in this document. The velocity fields with different equivalence ratio (0.7, 0.8 and 1) are presented. Near the burner a decrease in the equivalence ratio increases the injection velocity of the lateral jet and keeps a constant velocity in the central jet. For the turbulence intensity, near and far from the burner, an increase in the turbulence intensity is observed in the two layers of internal mixtures, this makes it possible to improve the mixing and increase the stability of the flame. Thus, there is an increase in the adiabatic temperature of the flame, which promotes heat transfer and improves thermal efficiency.

The use of hydrogen solves instability problems of the flame that are related to lean combustion, due to the high diffusivity and reactivity of hydrogen in combustion. The results showed that the addition of hydrogen increased the flame velocity and the temperature while reducing CO_2 and CO emissions due to the reduction of the carbon in the fuel.

Nomenclature

d	Tube internal tube, mm
m_{ng}	mass flow rate, $kg.s^{-1}$
ng	natural gas jet
P	thermal power, W
Re	Reynolds number
S	separation distance between the jets, mm
U_{ng}	nozzle exit velocity of gas, $m.s^{-1}$
U_{o2}	nozzle exit velocity of oxygen, $m.s^{-1}$
U	longitudinal mean velocity, $m.s^{-1}$
u’	longitudinal velocity fluctuation, $m.s^{-1}$
V	radial mean velocity, $m.s^{-1}$
v’	radial velocity fluctuation, $m.s^{-1}$
r, x	radial and longitudinal coordinate, mm

Greek symbols

μ	dynamic viscosity, $kg.m.s^{-1}$
ρ	gas density, $kg.m^{-3}$
α	percentage of hydrogen
Φ	transport terms
Φ	equivalence ratio
Γ_Φ	transport coefficient

Author details

Mohamed Ali Mergheni[1,2]*, Mohamed Mahdi Belhajbrahim[2], Toufik Boushaki[3] and Jean-Charles Sautet[4]

1 Department of Mechanical Engineering, College of Engineering, King Khalid University, Abha, Saudi Arabia

2 College of Engineering, University of Monastir, Monastir, Tunisia

3 ICARE-CNRS, Avenue de la Recherche Scientifique, University of Orléans, Orléans, France

4 CORIA, CNRS-Université et INSA de Rouen, Saint Etienne du Rouvray, Rouen, France

*Address all correspondence to: mmerghni@kku.edu.sa

References

[1] Das LM, Gulati R, Gupta PK. A comparative evaluation of the performance characteristics of a spark ignition engine using hydrogen and compressed natural gas as alternative fuels. International Journal of Hydrogen Energy. 2000;**25**:783-793

[2] Momirlan M, Veziroglu TN. The properties of hydrogen as fuel tomorrow in sustainable energy system for a cleaner planet. International Journal of Hydrogen Energy. 2005;**30**(7):795-802

[3] Genies B. L'oxydation une solution attractive pour minimiser les émissions de NOx des procédés industriels. Combustion et procédés industriels: Comment réduire les émissions de NOx, Rencontres et journées techniques. A.D. E.M.E; 1996. pp. 153-196

[4] Perthuis E. La Combustion Industrielle. Paris: Technip; 1983

[5] Krothapalli A, Bagadanoff D, Karamchetti K. Development and structure of a rectangular jet in a multiple jet configuration. AIAA Journal. 1980;**18**(8):945-950

[6] Moawad AK, Rajaratnam N, Stanley SJ. Mixing with multiple circular turbulent jets. Journal of Hydraulic Research. 2001;**39**(2):163-168

[7] Pani B, Dash R. Three dimensional single and multiple jets. Journal of Hydraulic Engineering. 1983;**10**:254-269

[8] Simonich JC. Isolated and interacting round parallel heated jets. AIAA Paper. 1986:86-0281

[9] Yimer I, Becker HA, Grandmaison EW. Development of flow from multiple jet burner. AIChE Journal. 1996;**74**:840-851

[10] Lee BJ, Kim JS, Lee S. Enhancement of blow-out limit by the interaction of multiple nonpremixed jet flames. Combustion Science and Technology. 2004;**176**:481-497

[11] Lenze B, Milano ME, Günther R. The mutual influence of multiple jet diffusion flames. Combustion Science and Technology. 1975;**11**:1-8

[12] Boushaki T, Mergheni MA, Sautet JC, Labegorre B. Effects of inclined jets on turbulent oxy-flame characteristics in a triple jet burner. Experimental Thermal and Fluid Science. 2008;**32**:1363-1370

[13] Faivre V, Poinsot T. Experimental and numerical investigations of jet active control for combustion applications. Journal of Turbulence. 2004;**5**:5-25

[14] Lesieur C. Modelisation de la combustion turbulente non premelangee dans un bruleur a jets separes: Application a la stabilisation d'une oxy-flamme [PhD thesis] INSA of Rouen; 2003

[15] Raghunatan S, Reid IM. A study of multiple jet. AIAA Paper. 1981;**19**: 124-127

[16] Salentey L. Etude expérimentale du comportement de brûleur à jets séparés: application à la recherche gaz naturel-oxygène pur [PhD thesis] University of Rouen; 2002

[17] Sautet JC, Salentey L, Ditaranto M, Samaniego JM. Length of natural gas-oxygen nonpremixed flames. Combustion Science and Technology. 2001;**166**:131-150

[18] Sautet JC, Boushaki T, Salentey L, Labegorre B. Oxy-combustion properties of interacting separated jets. Combustion Science and Technology. 2006;**178**:2075-2096

[19] BelhajBrahim MM, Mergheni MA, Ben Nasrallah S, Sautet JC. Numerical study of hydrogen enrichment effects in oxy-flame turbulent of three separated jets. Applied Thermal Engineering. 2017;**113**:490-498

[20] Mergheni MA, Boushaki T, Sautet JC, Ben Nasrallah S. Numerical study of oxy-flame characteristics in a burner with three separated jets. Applied Thermal Engineering. 2016;**111**: 49-98

[21] Yon S, Sautet JC. Flame lift-off height velocity flow and mixing of hythane in oxy-combustion in a burner with two separated jets. Applied Thermal Engineering. 2012;**32**:83-92

[22] Boushaki T, Sautet JC, Labegorre B. Control of flames by radial jet actuators in oxy-fuel burners. Combustion and Flame. 2009;**156**:2043-2055

Section 2

Investigations of Non-Conventional Engine Fuelling

Chapter 3

Effects of Biodiesel Blends Varied by Cetane Numbers and Oxygen Contents on Stationary Diesel Engine Performance and Exhaust Emissions

Semakula Maroa and Freddie Inambao

Abstract

This work investigated waste plastic pyrolysis oil (WPPO), 2-ethyl hexyl nitrate (EHN), and ethanol as sources of renewable energy, blending conventional diesel (CD), WPPO, and ethanol with EHN was to improve the combustion and performance characteristics of the WPPO blends. EHN has the potential to reduce emissions of CO, CO_2, UHC, NO_X, and PM. Ethanol improves viscosity, miscibility, and the oxygen content of WPPO. Mixing ratios were 50/WPPO25/E25, 60/WPPO20/E20, 70/WPPO15/E15, 80/WPPO10/E10, and 90/WPPO5/E5 for CD, waste plastic pyrolysis oil, and ethanol, respectively. The mixing ratio of EHN (0.01%) was based on the total quantity of blended fuel. Performance and emission characteristics of a stationary 4-cylinder water-cooled diesel Iveco power generator were evaluated with ASTM standards. At 1000 rpm, the BSFC was 0.043 kg/kWh compared to CD at 0.04 kg/kWh. Blend 90/WPPO5/E5 had the highest value of 14% for BTE, while the NO_X emissions for 90/WPPO5/E5, 80/WPPO10/E10, and 70/WPPO15/E15 were 384, 395, and 414 ppm, respectively, compared to CD fuel at 424 ppm. This is due to their densities of 792 kg/m^3, 825 kg/m^3 which are close to CD fuel at 845 kg/m^3 and the additive EHN. These results show blends of WPPO, ethanol and EHN reduce emissions, and improve engine performance, mimicking CD fuel.

Keywords: 2-ethyl hexyl nitrate, ethanol, oxygen content, ignition quality, waste plastic pyrolysis oil, cetane index

1. Introduction

Diesel engines the world over are the major power source in the automobile transport industry and nonroad powered engines. However, because of the issue of pollution associated with diesel exhaust, particularly particulate matter (PM) and nitrogen oxide (NO_X), there has been increasingly stringent regulation to control the manufacture and use of diesel engines. This has led to extensive research on improving diesel engines and fuel [1, 2]. The use of alternative fuels tops the list of measures to control diesel exhaust emissions as recommended by researcher [3]. Besides the use of alternative fuel to control and reduce emissions, other control strategies such as exhaust gas recirculation (EGR), diesel particulate filters (DPF), selective catalytic reduction (SCR), and catalytic converter combinations have been recommended but not as stand-alone technologies [4, 5].

The transport industry and nonroad diesel engines are major contributors to global gross domestic product. Nevertheless, their use affects human health and degrades the environment. The transport industry is responsible for one third of all environmental emissions of volatile organic compounds (VOCs), including two thirds of carbon monoxide (CO) emissions [6]. Carbon dioxide (CO_2) is a primary cause of global warming with 34 billion tons per year or 22% of all the global emissions per year [7], with a projected increase in 3% annually since 2011. This is projected to rise to 41 billion tons of CO_2 emissions by the year 2020 [8, 9]. Diesel engines release emissions, which lead to poor air quality, acid rain, smog, haze, and climate change. These factors increase the global disease burden due to respiratory system diseases and cancer [10].

The soluble organic fraction (SOF) and volatile organic fraction (VOF) are mainly due to exhaust dilution and the cooling process from fuel or evaporating lubricating oil, due to the process of oxidation. The control of VOC emissions is with high-pressure injection system catalytic converters and positive crankcase ventilation systems. The PM emissions of VOCs arising from evaporating lubrication oil and incomplete combustion have a combined emission rate of 0.06–2.2 g/bkWh for light diesel (LD) engines compared to heavy diesel (HD) engines at 0.5–1.5 g/bkWh [11, 12]. The condensation of oxidized and pyrolyzed products of fuel molecules is the leading cause for the formation of PM emissions composed of the nucleation and accumulation modes [13]. In emerging economies, air pollution is the leading cause of thousands of premature deaths estimated at 2.4 million annually by 2009 estimates [14]. Besides the usual toxics emitted by stationary and nonroad engines, diesel engines emit toxics such as formaldehyde, acrolein, acetaldehyde, and methanol. Exposure to these toxic emissions causes eye, skin, and mucous membrane irritation, besides affecting the nervous system. Therefore, the need for environmental protection has played a role in bringing together relevant stakeholders and government agencies. These agencies include the WHO, the Organization of Economic Cooperation and Development (OECD), the Inter-Governmental Panel on Climate Change (IPCC), the Environmental Protection Agency (EPA), the European Environmental Agency (EEA), and the International Energy Agency (IEA). For example, the USA government through the EPA has established the Reciprocating Internal Combustion Engine (RICE) rules, which cover stationary and nonroad engine emission regulations [15]. These rules are out of the scope of this chapter, but future work will discuss them in line with other European Union rules [16] and other global adopted emission regulations.

In order to meet modern requirements, diesel engines are designed with complex contrary goals to operate optimally in stationary and mobile operations. This requires high torque, low emissions, and high efficiency engines. For this reason, auxiliary diesel engine components such as turbochargers, EGR, and high-pressure injection systems are utilized today. These auxiliary parts are grouped into engine operating subsystems such as air, combustion, injection, and mechanical units to meet these operating demands. Since fuel is a major determinant in engine combustion and emission characteristics, the use of alternative fuel is being encouraged as a strategy to reduce emissions. The combustion of alternative fuel is different from the combustion of diesel, which is a fossil fuel, but they too cause emission problems as has been reported in a number of studies [17, 18]. To mitigate these problems, researchers have come up with combustion control strategies such as:

- homogeneous charge compression ignition (HCCI) [19, 20];

- pre-mixed charge compression ignition (PCCI) [21, 22];

- reactivity charge compression ignition (RCCI) [23]; and
- variant strategies to deal with emissions [24–30].

Modern day passenger vehicles and stationary engines are now evaluated using driving cycles such as the New European Driving Test Cycle (NEDC) and the Federal Test Procedure (FTP) mostly as bench operated chassis dynamometer tests [31]. However, it should be remembered that at engine start conditions, after-treatment techniques report poor performance as most of them operate with catalysts that are light-off temperature dependent. At ambient temperature, most catalysts cannot attain the light-off temperature when engines are started and operated. Since the year 2000 when EURO III was implemented, the NEDC procedure has been modified to eliminate the 40 s warm up before emission sampling can take place [32]. The new development initiative for diesel exhaust emission has already been established in the United States and Japan. The last decade has seen the European Union implementing similar standards and procedures, with the rest of the world expected to also implement changes as globalization and interdependency grows. A number of requirement have been implemented in the United States to nominally reduce 85–90% of NO_X, while for the Euro VI (2014), an additional reduction 65–70% of NO_X to match the US standards has been accepted as shown in **Figures 1** and **2** [33].

The combustion of diesel fuel depends on several factors that affect engine geometry, fuel properties, compression temperatures (especially of the combustion mixture), injection strategy applied, and the existing condition of the ambient temperatures as reported by the authors of Refs. [34, 35]. High cetane number additives together with the development of high volatility fuels [36, 37] have boosted diesel engine performance. The oxygenated additives in biodiesel blend components improve the combustion process, especially the octane rating. Additionally, oxygenated additives enhance and increase the cetane number. In other words, the oxygen in the additives supports the combustion of the fuel while at the same time reducing inert material such as NO_X formation in CI engines. These changes deal with the complexities of cold start, which impede engine starting at lower or subzero engine temperatures. Warm engines have a starting time delay of 1–2 s at ambient temperature conditions, compared to a low ambient temperature start-up time of 10 s [38, 39].

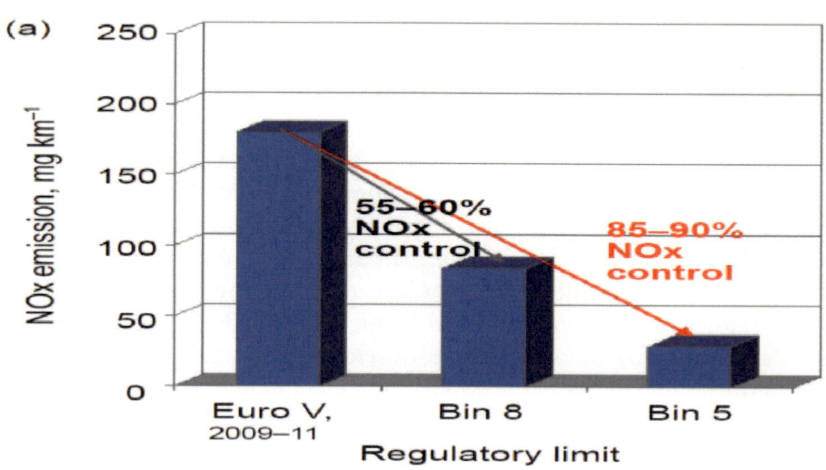

Figure 1.
Requirement to reduce about 55–60% of NO_X emissions for Euro V (2009) diesel to match the US Bin 8 maximum allowable emission in 45 US states [33].

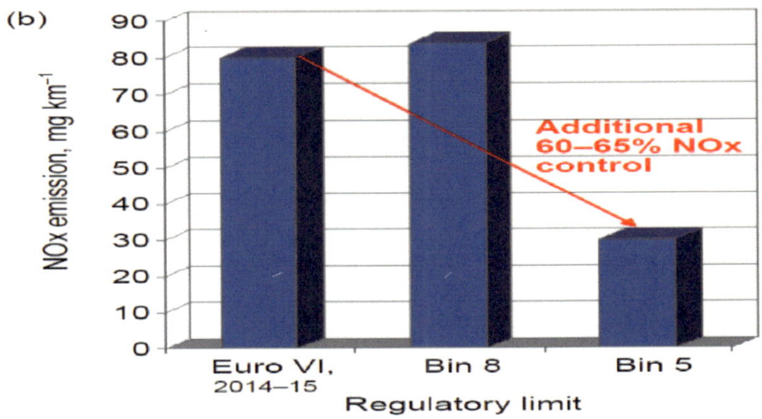

Figure 2.
Variation of NOx emission with the regulatory limit for 45 US states [33].

2. Regulatory development in diesel engine emission

Stringent diesel exhaust emission regulatory policies have been in operation in the United States and Japan since 2005. The European Union has also responded with additional emission regulatory standards called the EURO VI since 2014. Apart from these regulatory controls, there are market and political pressures on automobile manufacturers to continue to improve on efficiency while reducing emissions. These factors have been the driving force behind the significant technological progress in engine research and the transport industry in the past two decades.

2.1 Heavy duty diesel regulatory developments

The European Union commission on emissions in 2014 stipulated that the nominal NO_X emission limit must be 0.20 g/kWh^{-1} and the PM emission level must be 0.010 g/kWh^{-1}. This matched the US 2010 emission regulation, which put the emission limits at 0.26 g/kWh^{-1} for NO_X and 0.013 g/kWh^{-1} for PM emissions. The Japanese emissions regulation of 2009 stipulates 0.7 g/kWh^{-1} for NO_X emissions and 0.010 g/kWh^{-1} for PM emissions. However, it must be mentioned here that each of these countries propose a different transient testing cycle.

The European Union commission on pollution and emission has adopted a new world harmonized transient cycle (WHTC) that uses higher load and speed than the Japanese and American standards. Additionally, the European commission on emissions has set standards related to number-based PM standards with heavier in-use compliance measures as illustrated in **Figure 3**, by 2012. These measures are aimed at improving fuel economy and durability and lowering the cost of manufacturing and maintenance. The development in this segment is muted, mixed with conservatism and pragmatism. For example, the development in HD since 2004 has seen the US regulations matched and addressed through advanced EGR and intake charge boosting measures. However, later development starting from 2005 in Japan and 2007 in the United States has seen additional technologies added to cater for increased regulation. These two markets introduced diesel particulate filters (DPFs) to match the change in policy and regulation in the European Union with implementation of EURO V-VI emission regulations. This policy shift and regulation change has witnessed conventional engine technology adding the selective catalytic

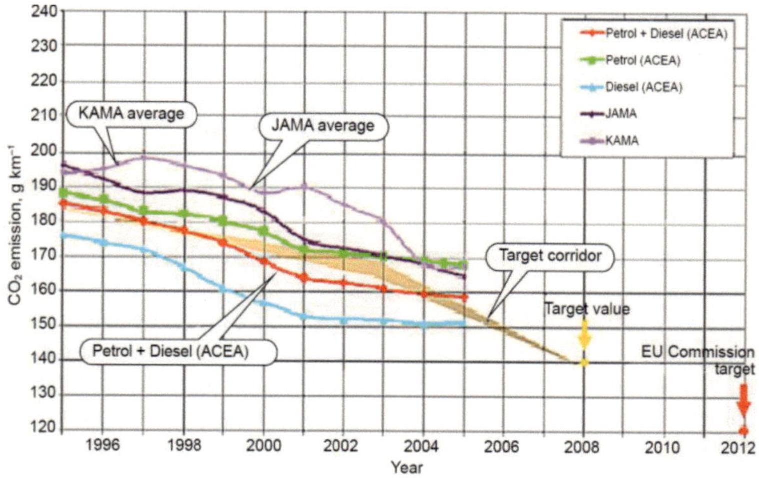

Figure 3.
Progress toward meeting the European Union voluntary CO_2 limits of the European Automobile Manufacturers Association (ACEA), Japan Automobile Manufacturers Association (JAMA), and Korea Automobile Manufacturers Association (KAMA) (courtesy of Daimler Chrysler) [40].

reduction (SCR) system in the fight against emission. Since 2009 and 2010, respectively, Japan and the United States have added significant incremental advances in emission compliant technologies, especially technologies that target low load emissions in HD engines. Researchers working on the traditional diesel combustion hardware and strategies are directing more effort to reduce LD engine category emissions with the future looking bright.

2.2 Light duty diesel regulatory development

Modern diesel engine development is driven by regulatory, market, and fuel efficiency demand. In addition, developments in spark ignition (SI) gasoline engines, electric vehicles, and new concepts in hybrid vehicles have had tremendous competitive pressure on diesel engine development, especially in the LD category. Diesel engine manufacturers are responding with the introduction of advanced fuel injection technology, exhaust gas recirculation (EGR) techniques,

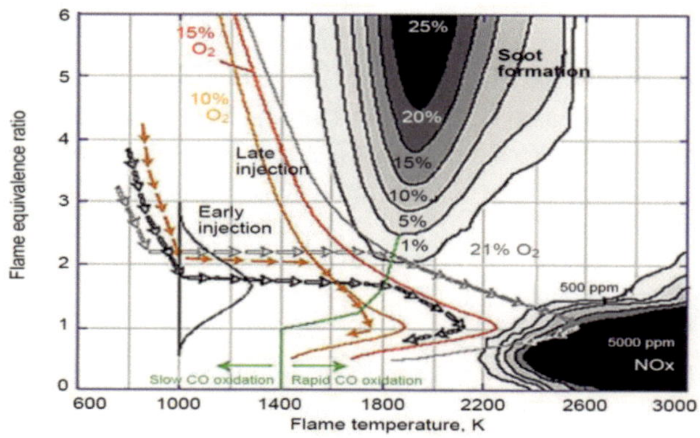

Figure 4.
Variation of flame equivalence ratio, temperature, and injection strategies and principles of advanced combustion (courtesy of Sandia National Laboratory) [44, 45].

two-stage turbocharging, variable valve actuation, closed loop combustion control, and advanced model-based controls. Development in advanced diesel engines has now achieved a specific output of 70 kW^{-1} and a brake mean effective pressure (BMEP) of 24 bars [41], hence meeting EURO VI emission standards [42, 43] as shown in **Figure 4**.

3. Control strategies for emissions in diesel engine

The world is now aware of the environmental and human health costs of pollution from diesel engines, which form the bulk of commercial and personal public transport systems. **Table 1** shows that there is an increase in the regulatory measures on toxic gas emissions. These regulations oblige vehicle manufacturers and transport industry service providers to be motivated to work harder to meet the appropriate standards and regulations. Among the techniques that have been employed to cut down on emissions are EGR, LNT, DOC, DPF, and SCR [46, 47]. However, there is no single method that meets emission standards by the regulatory bodies on vehicular emission.

3.1 Exhaust gas recirculation (EGR)

This is one of the most useful and successful techniques in the control of and fight against diesel exhaust emissions. EGR allows the recirculation of part of the diesel exhaust into the combustion chamber, to reburn together with the fresh intake charge [49] as shown in **Figure 5**.

This technology has been able to reduce NO_X emissions, but it causes an increase in UHC and CO emissions as compression temperatures decrease. It also affects engine thermal efficiency as shown in **Figure 6**. This technique has two methods for quantification of EGR flow rate, although there is no single method that is universally accepted. The two methods are the mass method and the gas concentration method [5]. These two methods are demonstrated in **Figure 5** and expressed in Eqs. (1) and (2):

$$r_{EGR} = \frac{\dot{m}_{EGR}}{\dot{m}_{air} + \dot{m}_f + \dot{m}_{EGR}} \quad (1)$$

$$\frac{[CO_2]_{int} - [CO_2]_{amb}}{[CO_2]_{exh} - [CO_2]_{amb}} \approx \frac{[CO_2]_{int}}{[CO_2]_{exh}} \quad (2)$$

where the $\dot{m}_{EGR}$ is the mass flow rate of the gas recirculated, $\dot{m}_{air}$ is the mass flow rate of fresh air, $\dot{m}_f$ is the mass flow rate of the injected fuel, and r_{EGR} is the mass fraction of the recirculated exhaust gases. $[CO_2]_{int}$ is the carbon dioxide at the intake side, $[CO_2]_{amb}$ is the ambient carbon dioxide, $[CO_2]_{exh}$ is the exhaust carbon dioxide (exit carbon dioxide).

3.2 The low NO_X trap (LNT)

This system is also known as NO_X storage reduction (NSR) and NO_X absorber catalyst (NAC). It has three main components, namely, the oxidation catalyst with platinum (Pt), the NO_X storage with barium (Ba), and the reduction catalyst with rhodium (Rh). The platinum catalyst is preferred as it reduces NO_X emissions at very low temperatures while offering a stable reaction in the presence of sulfur and H_2O [51, 52]. **Figure 7** shows the LNT three-stage catalytic process.

STD type	CO g/kWh	HC g/kWh	NO_X g/kWh	PM g/kWh
Euro I	4.5	1.1	8.0	0.61
Euro II	4	1.1	7.0	0.15
Euro III	2.1	0.66	5.0	0.13
Euro IV	1.5	0.46	3.5	0.02
Euro V	1.5	0.46	2.0	0.02
Euro VI	1.5	0.13	0.4	0.01

Table 1.
EURO standards for heavy-duty vehicles according to Delphi 2016–2017 as per Ref. [48]

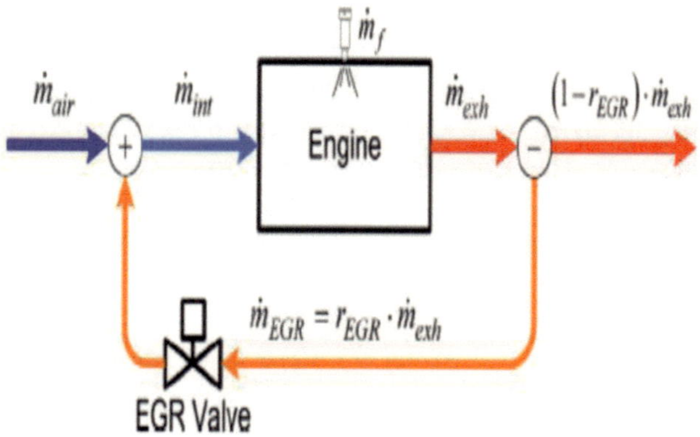

Figure 5.
EGR system nomenclature and control design for the EGR valve [4].

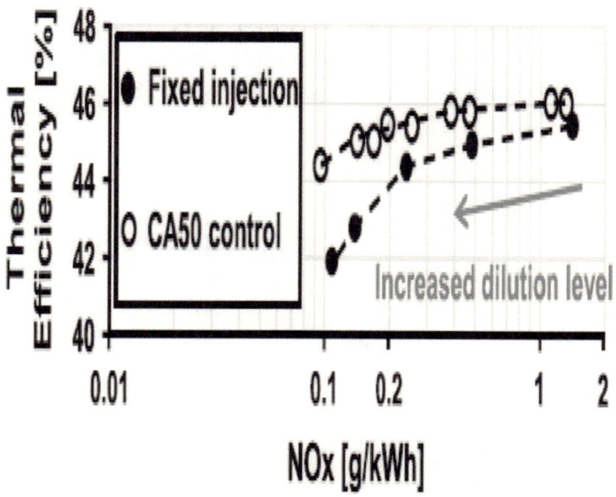

Figure 6.
Variation of engine thermal efficiency and NO_X with the influence of EGR dilution [50].

3.3 The selective catalyst reduction (SCR)

This is one of the most recent technology developments introduced for the control of diesel exhaust emissions. This system was originally introduced to cater for HD engines [53], but Audi and Volkswagen have also adopted it for their

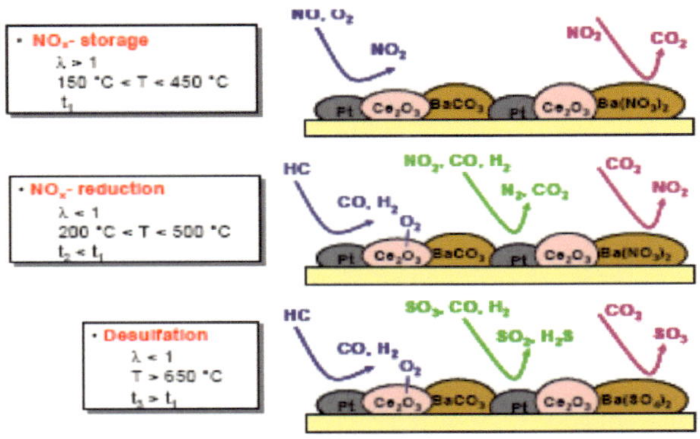

Figure 7.
The low NO_X trap (LNT) with three of its operating modes [53].

passenger vehicle and LD segments. The SCR system works by utilizing ammonia as a reductant in order to minimize NO_X emissions in the diesel exhaust by releasing N_2 and H_2O. This system therefore undergoes two processes during the working cycle, namely, hydrolysis and thermolysis as in Eqs. (3) and (4) for hydrolysis and thermolysis, respectively [54, 55].

$$HNCO + H_2O \rightarrow NH_3 + CO_2 \tag{3}$$

$$(NH_2)_2CO \rightarrow NH_3 + HNCO \tag{4}$$

In addition to the two processes of hydrolysis and thermolysis, SCR undergoes other chemical reactions to complete its normal cycle, thus reducing the emissions of NO_X further as in Eqs. (5)–(7). **Figure 8** shows a schematic diagram of an SCR system showing the oxidation catalyst, wall flow particulate filter, and the flow through the SCR catalyst. **Figure 8** also includes key components of a urea solution tank, a spray module, a static mixer, temperature, and NO_X sensor, courtesy of Robert Bosch GmbH [46].

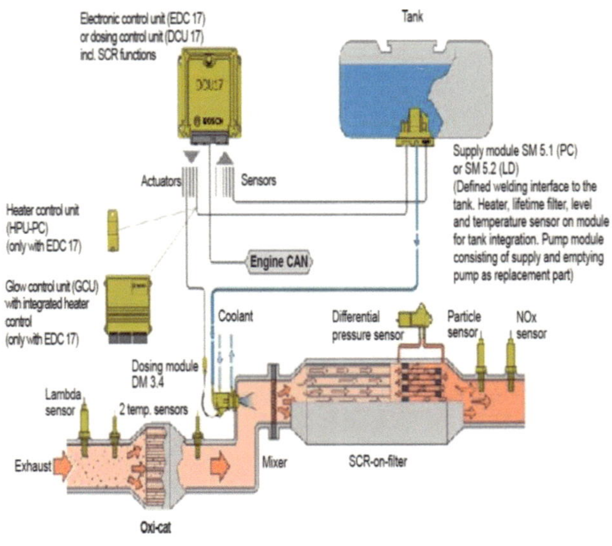

Figure 8.
Schematic diagram of the SCR NO_X control system as used in a standard production vehicle [46].

$$4NO + 4NH_3 + O_2 \rightarrow 4N_2 + 6H_2O \tag{5}$$

$$2NO + 2NO_2 + 4NH_3 \rightarrow 4N_2 + 6H_2O \tag{6}$$

$$6NO_2 + 8NH_3 \rightarrow 7N_2 + 12H_2O \tag{7}$$

3.4 Diesel particulate filter (DPF)

The DPF filter requires care to avoid excessive saturation and build-up of backpressure, both of which are harmful for engine operation and durability and increase fuel consumption and engine stress levels leading to premature failure of the filter and engine. DPF systems have been in operation in diesel exhaust emission control since the year 2000, primarily for removing PM emissions through physical filtration. DPFs are like a honeycomb with silicone carbide or cordierite written chemically as $2MgO - 2Al_2O_2 - 5SiO_2$. Both ends of the structure are blocked to force the particulate matter through the porous substrate walls, thus acting as a mechanical filtering system. These walls are made such that they offer little or no resistance to flow of exhaust gases while maintaining the power to collect particles [56] as shown in **Figure 9**.

3.5 Diesel oxidation catalyst (DOC)

The DOC is manufactured with the sole purpose of reducing CO and UHC emissions through oxidation of the hydrocarbons that are absorbed into the carbon particles. The DOC consists of a metal or a ceramic structure with an oxide mixture also called the wash coat that contains aluminum oxide (Al_2O_3), cerium oxide (CeO_2), zirconium oxide (ZrO_2), and an active metal catalyst of either platinum, palladium, or rhodium [54], as shown in **Figure 10**. For HD and LD vehicles in Europe, the United States, and Japan, the DOC is the after-treatment emission control systems of choice. The DOC with a platinum metal catalyst is the most popular among manufacturers and consumers. However, the DOC has the disadvantage of reacting with sulfur oxide and sulfur trioxide producing sulfates and sulfuric acid,

Figure 9.
Schematic of the working mechanism of a diesel particulate filter (DPF) [6].

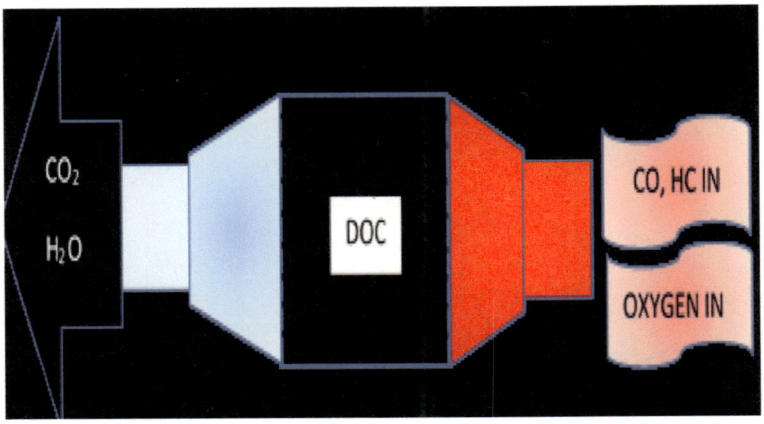

Figure 10.
Schematic diagram of a DOC and its operation in reducing emissions of CO and UHC through the process of oxidation [6].

which shortens the service life of the emission control system besides the additional effects on the natural environment and human health.

Six factors affect and influence the choice of a DOC filter:

- conversion factor;
- temperature stability;
- light-off temperature;
- tolerance to poisoning;
- cost of manufacturing the filter; and
- parametrical factors, including the density of the DOC filter measured in channels per square inch, the cross-sectional area, the channel wall thickness, and the length of the channels using the external dimensions [57, 58].

4. Methodology and experimental set-up

This experiment is making a case for blending of WPPO whose n-alkenes are lower by 25% in auto-ignition, compared to diesel fuel whose n-alkenes are good for auto-ignition. The aromatics, which affect PM emissions, are very low in WPPO blends. According to Refs. [59, 60], WPPO consists of iso-alkanes, n-alkanes, and olefins in the region of 27, 25, and 9%, respectively, with over 30% content being undefined due to complicated and complex chemical bond structures. However, aromatic cyclo-alkanes (naphthalene) and aromatics poor in auto-ignition were also found to be 40% by volume [61]. Blending was preferred to improve the low pour point to improve the cold starting characteristics of WPPO. Second, blending with ethanol was introduced to improve the fuel spray characteristics; ethanol is soluble and miscible in WPPO blends. Third, blending contributed to the reduction of the viscosity of WPPO biodiesel, thus further improving spray characteristics.

4.1 Engine tests

The experiment used a naturally aspirated four-cylinder diesel engine power generator, water cooled, direct injection, Iveco engine, in the Mechanical Engineering Department Laboratory, University of Kwazulu-Natal in Durban, South Africa. Using a defined flow rate of particles, PM emissions were detected by photoelectric measurement. Both the mass flow of the PM particles and the fuel were calculated as the sum of inlet air and fuel mass flow rate, and the result expressed in gram per kWh. To help in the analysis of the engine, pressure sensors and crankshaft position sensors and encoders were used. The aim of these sensors was to provide the in-cylinder pressure in relation to the crankshaft position variation.

Parameters	Position value
Ignition type	4 (Stroke)DICI
Number of cylinders	4 in-line
Cooling medium	Water
Manufacturer	Iveco
Revolutions per minute	2000
Brake power	43.40 kW @ 2000
Cylinder bore	104 mm
Piston stroke	115 mm
Compression ratio	17:1
Connecting-rod length	234
Engine capacity	2500 cc
Dynamometer make	234
Injection timing	12° bTDC
Maximum torque	206.9 Nm @ 1500
Injection pressure	250–272 Bar

Table 2.
Experimental engine specifications.

Property	Equipment	Standard
Kinematic viscosity	SVM 4000 (Anton Paar, UK)	ASTM D445
Flash point	NPM 550 (Norma lab, France)	ASTM D93
Oxidation stability	900 Rancimat (Metrohm, Switzerland)	ASTM D14112
CP/PP	NTE 500 (Norma lab, France)	ASTM D2500
Carbon residue	NMC 440 (Norma lab, France)	ASTM D4530
Total sulfur	5000 MULTI-EA (AJ Germany)	ASTM D5433
Calorific value	C 2500 basic calorimeter (IKA, UK)	ASTM D240
Density	SVM 3500 (Anton Paar, UK)	ASTM D1298
PM	AVL smoke meter	—
Gaseous emissions	Gas analyzer (MEXA 7000) Germany	—

Table 3.
List of equipment used in the experiment.

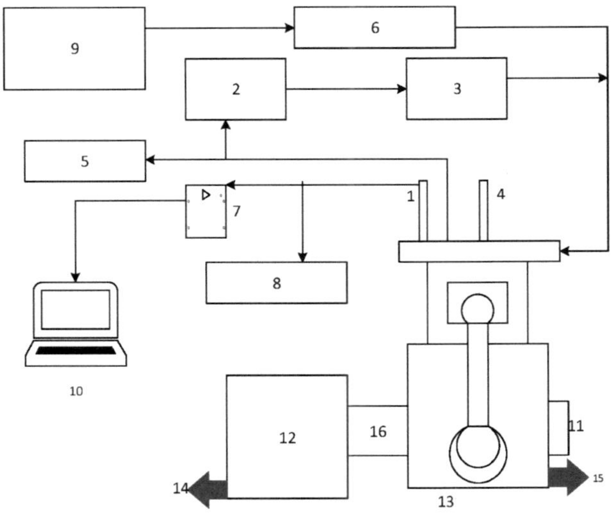

Figure 11.
Schematics of the test engine set up rig: (1) cylinder pressure sensor; (2) EGR control valve; (3) EGR cooler; (4) injection control unit; (5) exhaust gas exit; (6) air box; (7) signal amplifier; (8) gas analyzer; (9) air flow meter; (10) data acquisition system; (11) crank position sensor; (12) dynamometer; (13) engine; (14) cooling water exit from the dynamometer to the cooling tower; (15) cooling water exit from the engine to the cooling tower; and (16) dynamometer drive coupling.

The engine was coupled to a mechanical dynamometer with engine idling positions divided into two engine speed modes. The two speed modes were set at 500 and 1000 rpm as Mode 1, and Mode 2 as 1500 rpm and full load at 2000 rpm. The details of the engine and specifications and equipment are described in **Tables 2** and **3**. **Figure 11** shows a schematic of the engine test setup.

4.2 Physicochemical property analysis

WPPO by pyrolysis was obtained from a commercial plant whose production chart is shown in **Figure 12**. Ethanol, conventional diesel, and EHN were purchased

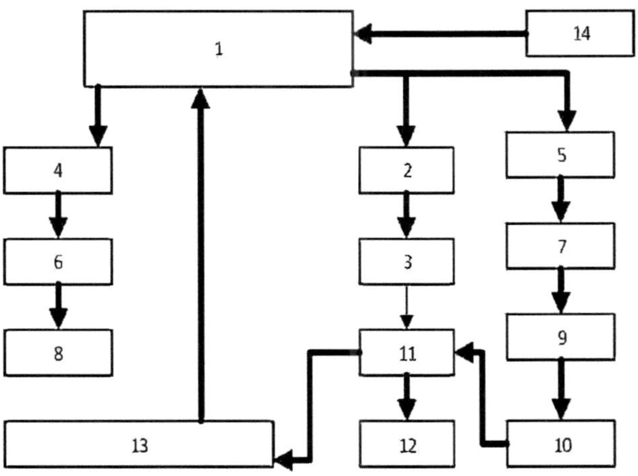

Figure 12.
Pyrolysis plant flow chart and its nomenclature: (1) pyrolysis reactor; (2) carbon black discharge; (3) carbon black deep processing; (4) exhaust smoke discharge; (5) gas separator; (6) smoke scrubber to take out color and odor; (7) condenser; (8) chimney; (9) oil tank; (10) synchronized gas purification; (11) synchronized gas-recycling system; (12) extra gas burning; (13) heating furnace during operation; and (14) loading of material.

Properties	Unit	CD	WPPO	Ethanol
Density @ 20°C	kg/m^3	845	825	792
Visc.@ 40°C	cSt	3.04	2.538	1.05
Cetane number	—	55	—	8.5
Flash point	°C	50	43	16
Fire point	°C	56	45	53
Carbon residue	%	22	0.015	—
Sulfur content	%	<0.028	0.248	—
Gross calories	kJ/kg	46500	43340	29700
Cetane index	–	46	65	—

Table 4.
Properties of diesel, WPPO, and ethanol before blending and addition of EHN.

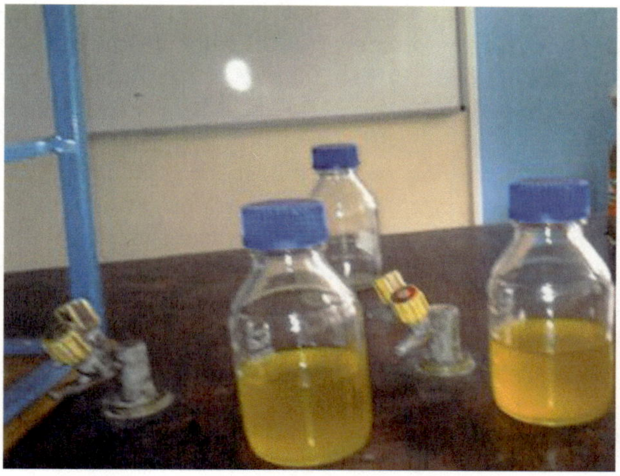

Figure 13.
The distillate samples from the waste plastic pyrolysis oil samples.

Property	Unit	CD	90/5/5	80/10/10	70/15/15	60/20/20	50/25/25	STANDARD
Density	Kg/m^3	845	838.5	834	830	825	823	ASTM D1298
Viscosity @ 40	cST	3.452	2.38	2.37	2.365	2.340	2.325	ASTM D445
Cetane number	-	45	59	62	64	65	69	ASTM D4737
GCV	kJ/kg	44840	41245	39985	38700	36800	34500	ASTM D4868
Sulfur content	%	<0.0124	0.0248	0.0249	0.0251	0.0253	0.0257	ASTM D4294
Oxygen	%	12.35	13.80	14.75	15.15	16.25	17.35	ASTM D5622
Carbon residue	%	74.85	75.35	76.40	77.55	78.25	79.65	ASTM D 7662
Flash point	°C	56.5	38.5	37.55	37.35	37.15	36.85	ASTM D93
Hydrogen	%	12.38	7.5	7.55	7.65	7.75	7.95	ASTM D7171

Table 5.
Properties of blended ratio mixtures of diesel, ethanol, WPPO with EHN.

from local outlets and blended using a homogenizer for 5 min at 3000 rpm. The properties of all samples were measured in the Chemical Engineering Laboratory of the University of Kwazulu-Natal in Durban, South Africa. **Table 3** shows some important physicochemical properties of the fuels before blending. **Table 4** shows physicochemical properties of fuels and their determined fuel properties after blending. **Figure 13** is a photograph of the sample distillates of WPPO obtained from pyrolysis. **Table 5** is showing properties of blended ratio mixtures of diesel, ethanol, WPPO with EHN.

5. Experimental results and discussion of diesel engine emissions

5.1 Brake-specific fuel consumption (BSFC)

Figure 14 is a variation in brake-specific fuel consumption (BSFC) with engine speed. The BSFC compared to the engine speed in **Figure 14** shows that as the speed increased, there is an equal increase of fuel consumed by the test engine. The values obtained at full engine speed (2000 rpm) for the blends of 90/WPPO5/E5, 80/WPPO10/E10, 70/WPPO15/E15, 60/WPPO20/E20, 50/WPPO25/E25, and CD were 0.04 kg/kWh, 0.041 g/kWh, 0.042 kg/kWh, 0.043 kg/kWh, and 0.035 kg/kWh, respectively.

At high engine speeds, the conversion of heat energy to mechanical energy increases with the increase in combustion temperature, leading to increased BSFC for the biodiesel; this increase is proportional to the difference in their heating values, which is identical to the findings of Ref. [62]. These blends of WPPO compare well to CD fuel and other biodiesel blends with comparative differences in the heating values.

However, from the graph, it is evident that as the blend ratio increases, there is a decrease in the BSFC across all the test fuels. Nevertheless, the values for all WPPO blends were slightly higher than the CD test fuel. The closeness of the values and the packed graph reveals a close resemblance and identical BSFC characteristics of WPPO, ethanol, and EHN compared to CD fuel. For example, at 500 rpm engine speed, the blend of 80/WPPO10/E10 had a value of 0.043 g/kWh compared to full engine speed (2000 rpm) with 0.041 kg/kWh, which is higher than CD test fuel with 0.04 kg/kWh at 1000-rpm engine speed and 0.035 kg/kWh at full engine speed (2000 rpm).

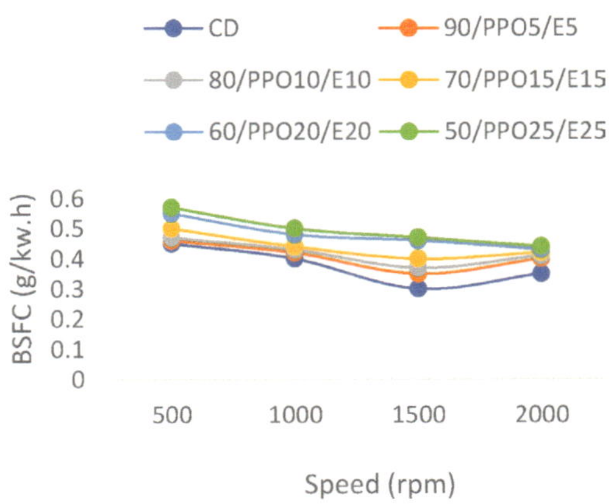

Figure 14.
BSFC versus engine speed.

5.2 Brake thermal efficiency (BTE)

The brake thermal efficiency (BTE) variations with engine speed are shown in **Figure 15**. The graphs show that as the speed increased, there was an increase in the BTE across all the test fuel blends of WPPO and CD up to 1500 rpm. At 1000 rpm engine speed, the values for blends 90/WPPO5/E5, 80/WPPO10/E10, 70/WPPO15/E15, 60/WPPO20/E20, 50/WPPO25/E25, and CD were 22, 21, 20, 18, 16.5, and 22.5%, respectively. As the blend ratio and engine speed increased, there was a decrease in the BTE within the WPPO blends but an increase in BTE across the blends. For example, at 500 rpm engine speed, 90/WPPO5/E5, 80/WPPO10/E10, 70/WPPO15/E15, 60/WPPO20/E20, and 50/WPPO25/E25 had values of 14, 13, 12.5, 11, and 9.5% compared to at 1000 rpm with values of 22, 21, 20, 18, and 16.5%, respectively.

The highest BTE value was 24% by blend 90/WPPO5/E5 at 1500-rpm engine speed compared to any other blend of WPPO, ethanol, and/or EHN. This could be due to the density, which is closer to CD, and the effect of blending, which improved this blend's physico-chemical properties. **Figure 15** shows values of 24.8, 23, 21, and 19%, respectively, for blends 80/WPPO10/E10, 70/WPPO15/E15, 60/WPPO20/E20, and 50/WPPO25/E25. Blend 50/WPPO25/E25 reported the lowest values compared to the other blends. At 500 rpm engine speed, the BTE value was 9.5% compared with full engine speed (2000 rpm) at 19%.

As the engine speed increased above 1500 rpm, the BTE suddenly dropped as the engine approached full engine speed (2000 rpm), as seen in **Figure 15**. There are a number of factors explaining the above results. For example, at this speed, there is a sudden drop of the air fuel ratio as the mixture becomes richer. This leads to incomplete combustion and heat release energy as more carbon molecules escape the combustion process. These increase the dissociation heat losses by the engine, hence a fall in BTE. Additionally, decreased BTE with biodiesel blends could be due to their low calorific value, higher viscosity, high volatility, and poor spray characteristics. These findings are consistent with other studies by the authors of Refs. [63–65].

5.3 Unburnt hydrocarbon (UHC) concentration

Unburnt hydrocarbon (UHC) concentrations largely indicate the quality of the combustion in an internal combustion engine. UHC concentrations are formed from vaporized unburnt hydrocarbon fuel and partially burnt fuel by-products exiting the combustion chamber diesel exhaust system. UHC concentrations are inherently independent of the air fuel ratio of any working engine [6].

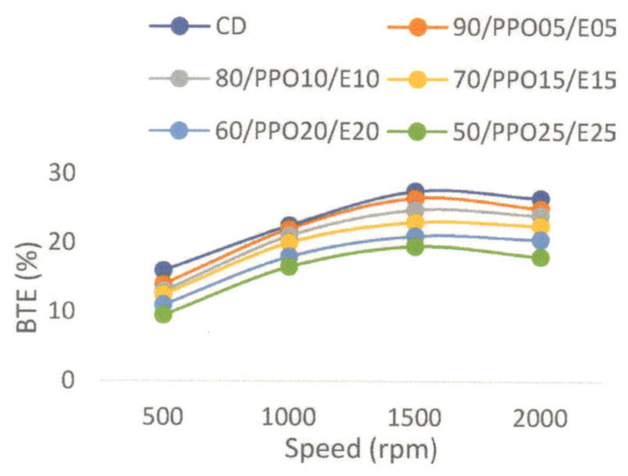

Figure 15.
Brake thermal efficiency versus engine speed.

In compression ignition (CI) engines, UHC concentrations are due to insufficient temperature, especially around the cylinder walls or in pockets. UHC concentrations are also formed through system malfunction, especially in input data failure in modern fuel injection systems. The higher hydrocarbon concentrations may be due to hydrogen radicals in the diesel-ethanol-WPPO-EHN blends. Principally, these concentrations are prevalent during light loads, when the combustion mixture is lean. This period is marked by a lower fuel ratio making the lean fuel-air mixture the primary source of the light load concentrations because of the lack of completion of the combustion during normal engine operating cycles. Hydrocarbon concentrations are not limited to vehicle exhaust systems but can occur in the entire vehicle fuel system from vapors during dispensing and distribution of fuel, which accounts for 15–20%, with the crankcase providing 20–30%. However, diesel exhaust remains the main culprit in engine emissions accounting for 50–60% of all the UHC concentration [66, 67].

Figure 16 shows the variation of UHC emission with engine speed in the stationary diesel power generator using blends of biodiesel. As the engine speed was increased, the UHC concentration increased too. However, the increase was more significant as the engine speed was in intermediate speeds of 1500 rpm moving to or approaching full engine speed (2000 rpm). For example, at 1000 rpm, the values of blends 90/WPPO5/E5, 80/WPPO10/E10, 70/WPPO15/E15, 60/WPPO20/E20, and 50/WPPO25/E25 were 22, 21, 20, 18, and 15 ppm, respectively, compared to full engine speed (2000 rpm) with 35, 34, 32, 29, and 26 ppm. This leads to the conclusion that at high engine speeds, the values of UHC concentration is significantly high for all the blends of WPPO, ethanol, and EHN, although still comparatively low compared to CD fuel.

The UHC concentration from the blends 90/WPPO5/E5 and 80/WPPO10/E10 had higher values although from the graph plot in **Figure 16**, the values are still low compared to the values of CD test fuel. However, the general trend reported by the graph in **Figure 16** shows that as the blend ratio increased, there was a significant reduction in the UHC concentration, observed across all the test fuels irrespective of the engine speed condition, for all the blends tested compared to CD fuel. The reduction in UHC concentration is attributed to the high oxygen content and cetane number of the blends. The high oxygen content supports combustion, while the high cetane number reduces ignition delay. This is identical to other studies by other researchers [68–72].

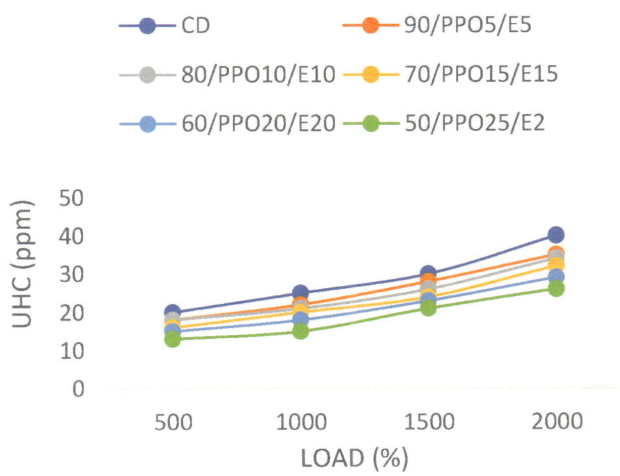

Figure 16.
Unburnt hydrocarbons versus engine speed.

The high fraction of ethanol in blends 70/WPPO15/E15, 60/WPPO20/E20, and 50/WPPO25/E25 contributed to the increase in the concentration of UHC, which is identical to the findings of Refs. [73, 74] who observed it in SI engine cylinder walls, crevices, and quenched cylinder walls, especially when richer air-alcohol mixtures were introduced. This type of UHC depends on the following factors: engine adjustments, engine design, and the type of fuel used in an engine. However, the engine-operating environment can sometimes contribute to the type of UHC concentration produced. This is observed especially when the temperature range is 400–600°C in the combustion chamber. At this temperature range, the hydrocarbons continue to experience reaction in the diesel exhaust pipe, thus lowering or increasing the concentration of the UHC in the exiting exhaust gas [75, 76].

5.4 Carbon monoxide (CO) formation

CO concentrations are a direct result of incomplete combustion, which results from hydrocarbons due to the failure of oxidation in the combustion process in diesel engines. This is true especially where the excess air factor λ meets the conditions $\lambda < 1$ for SI engines. Carbon monoxide is a colorless, tasteless, and odorless toxic gas, which is primarily a product of incomplete combustion of carbon containing fuels [6]. The United States is the single largest producer of carbon monoxide from anthropogenic sources as shown in **Figure 17** [77]. Carbon oxidation mechanisms are mostly determined by the equivalence ratio. Carbon monoxide concentrations mainly form in the areas of heavy traffic, parking garages, and under buildings, overheads, and overhangs. CO health effects include headaches and dizziness, but extreme exposure can lead to death.

Figure 18 is the variation of CO with engine speed in a stationary diesel power generator. The graph reveals that as the engine speed and the blend ratio increased 90/WPPO5/E5, 80/WPPO10/E10, 70/WPPO15/E15, 60/WPPO20/E20, and 50/WPPO25/E25, the CO concentration decreased up to 1500 rpm of engine speed. Thereafter, the blends reported a continuous increase as the engine speed was approaching full engine speed (2000 rpm). At 500-rpm engine speed, the blends of 90/WPPO5/E5, 80/WPPO10/E10, 70/WPPO15/E15, 60/WPPO20/E20, and 50/WPPO25/E25 reported values of 0.055, 0.0565, 0.06, 0.0615, and 0.0625%.

However, as the speed is increased to 1500 rpm, the values were 0.035, 0.0375, 0.0445, and 0.0475%, respectively. At full engine speed (2000 rpm), all the test fuels showed increased CO concentration with blends 90/WPPO5/E5 and 80/

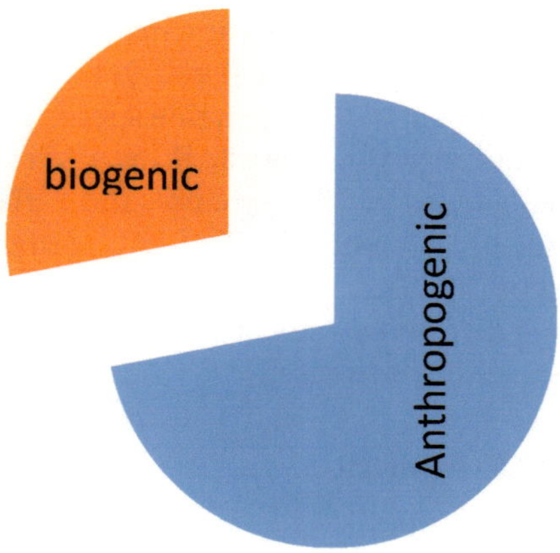

Figure 17.
Carbon monoxide (CO) concentrations by anthropogenic and biogenic sources in the United States [77].

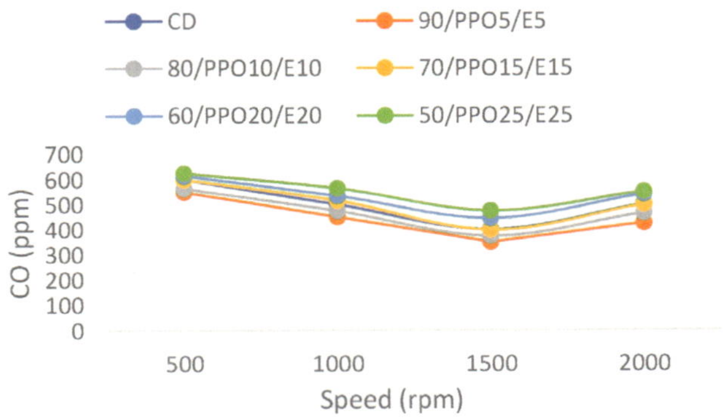

Figure 18.
Carbon monoxide versus engine speed.

WPPO10/E10 reporting the lowest concentration among the test blends across all the engine speed conditions. At 1000 rpm, the blends reported values of 0.0445 and 0.0475% compared to full engine speed (2000 rpm) with 0.0425 and 0.0465%, respectively. The increased CO concentration, although lower than diesel fuel, can be attributed to partial combustion [78] as the speed increased and the presence of ethanol, which shortened ignition delay, thus increasing CO concentration.

As the engine speed and the blend ratio increased, there was an increase in the CO emission across all the engine speeds and within the blends and CD test fuel. At 1000 rpm engine speed, the values of the blends and CD were 0.045, 0.0475, 0.0515, 0.0535, 0.0565, and 0.05% for 90/WPPO5/E5, 80/WPPO10/E10, 70/WPPO15/E15, 60/WPPO20/E20, 50/WPPO25/E25, and CD, respectively. The above values obtained from **Figure 18** suggest that there was a reduction in CO concentration across all test fuels irrespective of blend ratio and type of fuel except at high engine speeds exceeding 1500 rpm to full engine speed (2000 rpm). After this point, there was a steady increase in the concentration of CO.

CO concentration is a direct result of poor oxidation of the hydrocarbon fuels in the combustion chamber but is determined by the local fuel/air equivalence ratio. The above scenario is due to the air/fuel ratio becoming richer as the speed increased, leading to insufficient mixing of oxygen and fuel molecules. Compared to CD, all the biodiesels tested showed decreased CO concentration due to the high oxygen content in the test biodiesels and the addition of EHN, which greatly increased the cetane number (CN). This is identical to the studies by the authors of Refs. [79, 80]. The initial concentrations were greater at the starting engine speed of 500 rpm due to low temperature and emission instability processes at lower engine speeds, which are identical to the studies of Ref. [81]. However, as the engine speed increased from 1500 rpm toward full engine speed (2000 rpm), there was an observed increase in CO concentration, despite the oxygen content of the biodiesel and increased CN of the blends of WPPO, ethanol, and EHN. This disagreement in experimental results is due to differences in CN for the different biodiesel test fuel blends used. The increment in CN as the blend ratio increased led to increases in fuel quantity burnt during diffusive combustion, hence increasing CO concentration as the quality of combustion decreased.

5.5 Particulate matter (PM) formation

PM is agglomerates of small particle phase compounds resulting from the combustion of partially burned lubrication oil, the ash content from the fuel, sulfates from the engine cylinder wall, lubrication oil, and water from condensation and the combustion process [82]. These emitted compounds comprise elemental carbon (EC), organic carbon (OC) trace, and unknown compounds. Both EC and OC contribute to the toxicity of PM, regional haze, and climate change; therefore, PM concentration negatively affects the environment and human health [83]. The Global Burden of Disease Index reports that these types of emission are now responsible for 3.2 million deaths due to $PM_{2.5}$ ambient pollution [84]. Besides this, PM concentration causes deposit formation in the combustion chamber, fouling of emission control systems such as EGR and DPF and increased engine wear and premature failure.

PM concentration is primarily controlled by factors such as fuel quality (sulfur and ash content in fuel), engine lubrication oil quality, fuel consumption per combustion cycle of the engine, exhaust cooling rate, and the combustion process or strategy applied [85]. A number of PM characterization research works have been conducted categorizing PM concentration as 41% carbon, 7% unburned fuel, 25% unburned oil, 14% sulfates, water, 13% ash, and other concentrations [83]. However, an earlier study conducted by Agrawal et al. [86] reported that particulate concentration contains ≅31% elemental carbon, ≅14% sulfates and moisture, ≅7% unburnt fuel, and ≅40% unburnt lubricating oil. A study by Thiruvengadam et al. [87] yielded a similar outcome in terms of PM concentration except that the study was based on natural gas engine technology.

PM concentration is divided into three main components: SOF, soot, and inorganic fraction (IF), 50% of which is released as soot in the diesel exhaust pipe. SOF emissions are made up of condensed hydrocarbons embedded within the soot emissions in the form of very fine particles. The size distribution of PM concentration has three peaks: the nucleation peak, which includes all volatile hydrocarbons (D_p<~30 nm), the accumulation mode (~30 nm< D_p<~500 nm), and the coarse mode (~500 nm<D_p<~10 μm) [88]. These emissions are more pronounced during starting and engine idling when engine temperatures are reportedly very low [89].

Studies on OC/EC in PM samples show that their ratio is elevated in biodiesel combustion as the biodiesel blend ratio increases. This is mainly due to the high oxygen content in biodiesel and plays a major role in the generation of soot particles and final oxidation. For example, in a study by Chuepeng et al. [90], the authors reported that the OC fraction for B30 was greater than ULSD regardless of the engine speed and operating conditions. In another study by Williams et al. [91], a similar pattern was established for OC and EC as B100 > B20 > Diesel. This is identical to the studies of Ref. [90], which suggested an increased OC content with increased biodiesel fraction in a blend.

Cheung et al. [92] used soy methyl esters in an LD engine and found that the EC fraction was lower than during diesel operation. Nevertheless, the OC fraction in the PM concentration sample became identical for both LD and HD engines with the New European Driving Cycle (NEDC). However, a study by Song et al. [10] differs with this finding. Using cottonseed biodiesel, the authors reported decreased OC and EC driving conditions. This was mainly due to engine operating conditions, test methods, and test fuel chemical properties [83]. However, these studies have been inconsistent and inconclusive in the literature surveyed. For example, this is revealed in the studies of Refs. [90, 92–96].

DPF filters have now become part of virtually all diesel vehicles in the leading industrialized countries in the world (Europe, the United States, and Japan). DPF filters have had a high market penetration in Japanese and American LD and HD trucks since 2007. For smaller vehicle applications, subsequent developments have incorporated the diesel oxidation catalyst (DOC) function into the filter as reported by the authors of Refs. [10, 97]. It should be noted that for PM emission control in medium engines, the methods and approaches used are similar to the LD engines. However, in the US market, auxiliary fuel injectors and burners are incorporated into the diesel exhaust to regenerate DPFs. This method has concerns over oil dilution in the crankcase and requires a separation with the engine management system demands, so it has become more complex in the manner of its development and use [98, 99].

Advances in the science of materials have greatly increased and therefore influenced the development in filter materials for LD and HD engines. LD vehicles have seen silicon carbide types of filters becoming standard installation, although the alternative use of aluminum titanate is gradually replacing it [97]. However, aided by better engine controls, the industry has now moved to cordierite filters [101, 102]. **Figure 19** shows new hybrid developments in DPF filtering technology, which reduces 95% of NO_X that comes from the DPF filter.

As shown in **Figure 20**, speed affects particle emission of blends. Nevertheless, differences in engine operating conditions, particulate formation, in-cylinder combustion processes, and engine type give mixed results and conclusions in PM emission studies. In **Figure 20**, it is evident that as speed increases, combustion time (residence time) is reduced, which reduces the reoxidation and combustion of the constituents of the process. This aptly explains the reason behind increased PM particle size and concentration as the speed tends toward full engine speed (2000 rpm), as typified in the graph in **Figure 20**. For example, PM concentration at 500 rpm is 0.15, 0.11, 0.094, 0.086, 0.063, and 0.051 kg/kWh, respectively, for CD, 90/WPPO5/E5, 80/WPPO10/E10, 70/WPPO15/E15, 60/WPPO20/E20, and 50/WPPO25/E25.

However, as the speed increases from 500 to 1500 rpm, which is an intermediate speed, the PM emission increases and almost doubles to 0.29, 0.25, 0.235, 0.213, 0.183, and 0.57 g/kWh. These are for CD, 90/WPPO5/E5, 80/WPPO10/E10, 70/WPPO15/E15, 60/WPPO20/E20, and 50/WPPO25/E25, respectively. These findings are identical to the studies of Refs. [103, 104]. In other words, these blends, when combusting, produce low heat loss to the wall resulting in increased soot oxidation, which is also reported conclusively in a study by Di Iorio et al. [105] and is identical to the findings of this work in **Figure 20**.

Since PM concentrations are influenced by engine operating conditions at 1500–2000 rpm, PM concentration decreases with increased blend fraction. The reduction is more with higher blend ratios 70/WPPO15/E15, 60/WPPO20/E20, and 50/WPPO25/E25 at 2000 rpm. However, there is a reversed reduction in CD fuel compared to the blends of WPPO as shown in **Figure 20**. This is due to diffusive combustion as the blend ratio increases (tends to B100) and the oxygen content of the blends increases. These findings are identical to the findings of a study by Di Iorio et al. [105].

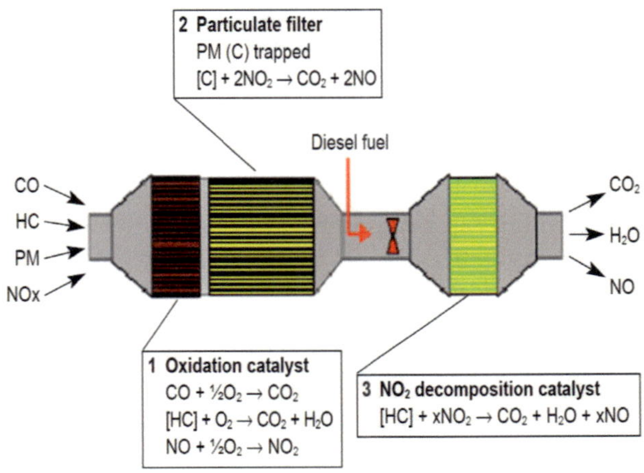

Figure 19.
A new NO_2 remediation system reduces 95% of the NO_2 emissions from catalyzed filter systems (courtesy of Technical University Dresden and Johnson Matthey) [100].

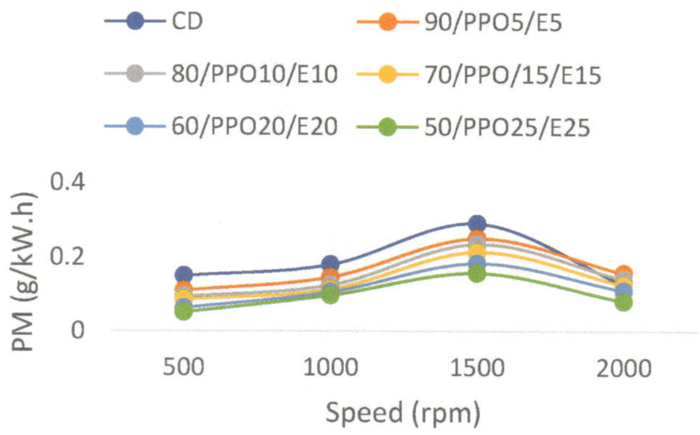

Figure 20.
PM emission for different blends of WPPO biodiesel fuel from 500 rpm to full engine speed (2000 rpm).

5.6 Carbon dioxide (CO_2) concentration

CO_2 is one of the gases responsible for maintaining the earth's optimal ecosystem balance. It enriches plants through the photosynthesis process and provides other benefits for the environment. However, CO_2 has become a topical global issue in recent decades due to its increase from levels of 0.04% in the atmosphere. The increase in CO_2 causes an increase in global temperatures due to the effect of blanketing. There are generally two sources of CO_2 formation: human activities and naturally occurring sources such as the ocean-atmosphere exchange, plant and animal respiration, soil respiration, decomposition of waste and elements, and volcanic eruptions. The majority of the human sources are due to the burning of hydrocarbon fuels in transport and power generation, land activities such as mining and agriculture, and industrial processes and manufacturing. The main gas produced from human activity is greenhouse gas associated with activities such as combustion of fossil fuels, namely, coal, natural gas, and oil for commercial and transportation services [106].

Figure 21 shows the variation of CO_2 with engine speed. The graph shows that as the blend ratio and engine speed increased, CO_2 concentration increased, but compared to CD, their emission levels were still lower and almost identical. At 500 rpm engine speed, the values of CD and the blends of 90/WPPO5/E5, 80/WPPO10/E10, 70/WPPO15/E15, 60/WPPO20/E20, and 50/WPPO25/E25 were 3.58, 3.35, 2.95, 2.6, 2.55, and 2.25%, respectively.

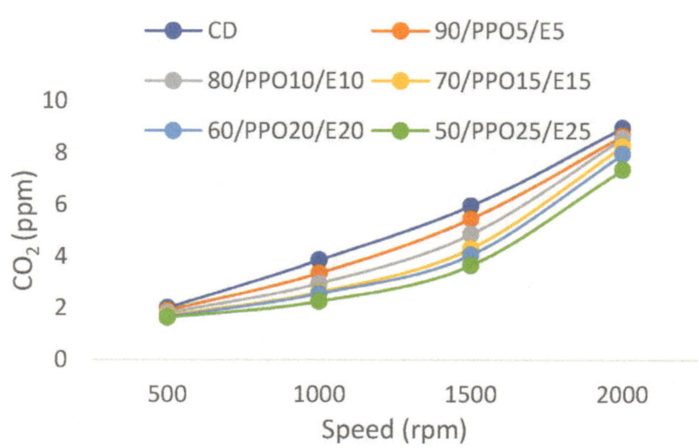

Figure 21.
CO_2 versus engine speed.

Figure 21 also shows that as the speed increased, there was a significant increase in the CO_2 concentration across all test fuels, although with lower values as the blend ratio increased. For example, CD fuel had values of 2, 3.85, 5.95, and 8.95% for engine speeds of 500, 1000, 1500, and 2000 rpm compared to blend 80/WPPO10/E10 with 1.8, 2.95, 4.85, and 8.55% for similar speeds. The blend with the lowest value of CO_2 emission was 50/WPPO25/E25 with values of 1.62, 2.25, 3.65, and 7.35%, respectively, for engine speeds of 500, 1000, 1500, and 2000 rpm, respectively. The increased carbon concentration in biodiesel blends is due to the reduction in the quantity of carbon relative to the increased oxygen ratio. However, the lower CO_2 concentration levels in comparison to CD fuel are due to factors explained under BTE and the equal energy balance generated by the addition of alcohol.

5.7 Nitrogen oxide (NO_X) concentration

NO_X concentration and its oxidized product NO_2 are the primary preserve of the diesel engine, constituting 85–95% of the total emission of a diesel engine. There are two fundamental differences between the two gases: whereas NO_X is odorless and colorless, NO_2 is reddish with a pungent smell [107]. It should be mentioned here that NO_2 is five times more toxic than NO_X gas and is a health hazard to the human respiratory system. It irritates the respiratory system and lowers the resistance to diseases such as the common cold and influenza [9, 108].

SCR is one of the leading NO_X emission control techniques for both LD and HD vehicles. This system entered the market in Japan and Europe for the HD category in 2005 compared to the US market in 2010. In the Japanese market and in Europe, zeolite and vanadium-based catalysts are utilized, respectively. The zeolite SCR catalyst combination performs better and has higher temperature tolerance levels. There is ongoing research to improve low temperature performance for more accurate NO_2 and NO_X concentration predictions [110–112].

The low NO_X trap (LNT) is a cheaper option for engines that are 2000–2500 cc [113, 114]. This type of emission control technique works better with mixed-mode engines to reduce low-load NO_X that is a persistent problem in SCR systems. This allows the LNT to focus on high temperature NO_X that is entering at temperatures over 300°C, thus eliminating between 60 and 70% of the platinum group metals (PGMs) [115]. This makes the LNT technology cheaper and economically appealing to the LD engine classification of 5000–6000 cc capacity [116, 117]. However, for medium- and heavy-duty vehicles, high temperature solutions have been developed to address the challenge of high load requirements of the US NTE regulatory condition as reported by the authors of Refs. [101, 118].

The LNT technique suffers due to contamination from sulfur, which shortens and affects its service life and durability. Earlier versions of LNT lost 50% filtration capacity, while the current generation of LNTs loses only 25% [119, 120]. Desulfication can be accomplished by passing a rich hot steam of diesel fuel at 700°C for 10 min at service intervals of 5000–10,000 km. **Figure 22** shows a new concept of combining the SCR emission control system with the LNT emission control system.

NO_X concentration is now known to be temperature dependent due to their equilibrium concentration presence in the combustion chamber. NO_X when mixed in high temperature adiabatically in the temperature range of 2000–3000 k forms NO_X concentration, which is then exited through the diesel exhaust system [121]. The NO_X concentration has four basic mechanisms of formation within the combustion chamber of a diesel engine: the Zeldovich mechanism also called the thermal NO_X route, the prompt mechanism, the fuel mechanism, and the NNH mechanism [122]. The variation of engine speed with NO_X concentration is shown in **Figure 23**. The graph shows that as the engine speed was increased, there was an increase in the

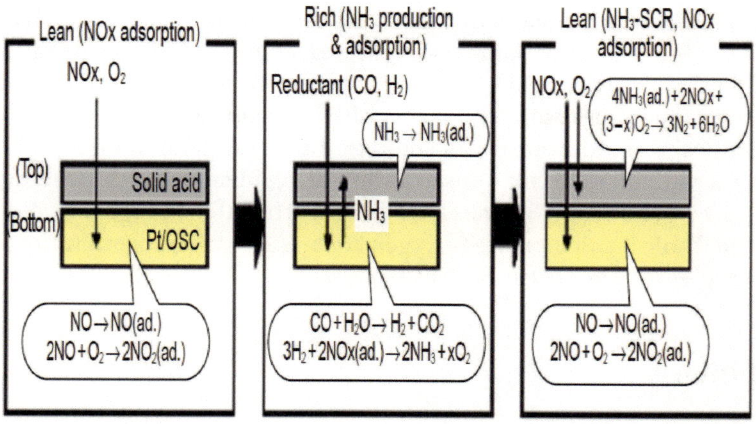

Figure 22.
The concept of employing a NO_X absorber with a double SCR layer configuration [109].

NO_X concentration irrespective of fuel, blend ratio, or additive. However, the value of NO_X concentration from the blends 90/WPPO5/E5, 80/WPPO10/E10, and 70/WPPO15/E15 reported lower values than CD fuel. For example, at 1000 rpm, the value of the blends was 385, 396, and 415 ppm, compared to CD fuel at 425 ppm.

Blends 60/WPPO20/E20 and 50/WPPO25/E25 had the highest NO_X concentration compared to the other blends of 90/WPPO5/E5, 80/WPPO10/E10, and 70/WPPO15/E15 across all the engine speed conditions tested. At 500 rpm engine speed, the two blends had values of 205 and 200 ppm, respectively. At full engine speed (2000 rpm), NO_X concentration values increased to 925 and 885 ppm compared to blend 90/WPPO5/E5 at 197 ppm and 792 ppm at full engine speed (2000 rpm). The graph in **Figure 23** shows that as the blend ratio increased, there was a direct increase in the concentration of NO_X across all the blended test fuels. However, blend 90/WPPO5/E5 reported the lowest values of NO_X concentration compared to all the other blends. The formation of NO_X in biodiesel combustion depends on the combustion temperatures and combustion zone oxygen concentration. With high blend ratios of 70/WPPO15/E15, 60/WPPO20/E20, and 50/WPPO25/E25, the combustion process is shortened, thus leading to failure to provide enough cooling effect to decrease peak combustion temperatures leading to increased NO_X.

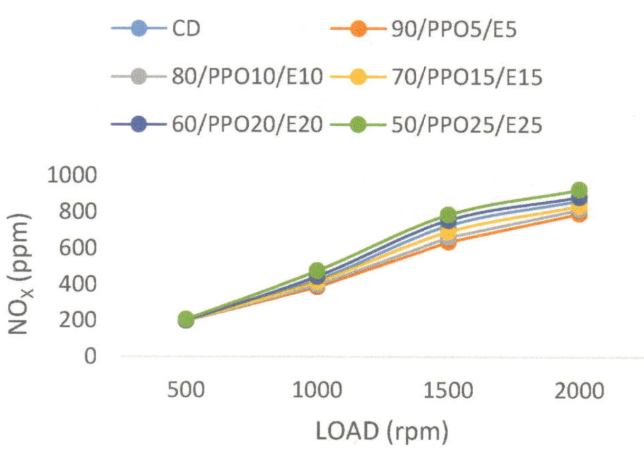

Figure 23.
Oxides of nitrogen versus engine speed.

These findings seem to show that there is a correlation between the alcohol content in the fuel and peak flame temperatures, content of nitrogen, and oxygen availability [123]. Increased NO_X concentration is attributed to the presence of nitrogen from the cetane number improver ENH and other contaminants from the WPPO composition. Additionally, it could be due to the generation of radicals of hydrocarbon through molecular unsaturation being identical to the findings of Refs. [124, 125]. However, the NO_X levels are still low, attributed to high CNs of the tested biodiesels in **Table 3** and increased oxygen content due to the blend ratios. These findings are identical to the findings of Ref. [126].

6. Conclusion

- This study thus makes a strong case for alternative fuels to replace petroleum-based fossil fuels like diesel commonly used as the primary propulsion fuel in transport and power generation. This work looks at the concept of waste to energy and waste resource utilization in an era when environmental concerns and awareness are at the pick of development agenda across the globe.

- Lower blend ratios 90/WPPO5/E5 and 80/WPPO10/E10 exhibit identical brake-specific fuel consumption (BSFC) of conventional diesel test fuel compared to the other blends. These blends show the lowest BSFC values compared to the others.

- The brake thermal efficiency of blend 90/WPPO5/E5 (90% conventional diesel, waste plastic pyrolysis oil 5%, an ethanol 5% by volume) showed values, which were very close to the values of conventional diesel fuel values. This was attributed to close density values and the gross calorific values of waste plastic pyrolysis oil (WPPO) blends, which showed marginal differences. This case was apparent especially at lower blend ratios of all the mixtures and blends tested.

- There was a reduction in unburnt hydrocarbon (UHC) concentration with the use of WPPO blends, ethanol, and 2-ethyl hexyl nitrate (EHN), with a notable reduction in oxides of nitrogen concentration especially for the blend 90/WPPO5/E (90% conventional diesel, waste plastic pyrolysis oil 5%, and ethanol 5% by volume). This was a clear indication that this blend performed well when compared with petroleum conventional diesel.

- Although there was indicated increase in the concentration of CO, CO_2 NO_X and UHC, for all the blends of WPPO, ethanol and EHN. There was a clear indication that the emission levels were notably lower than the emission levels of conventional petroleum diesel, based on the ASTM measurements used in this study. However, when comparisons for overall values of concentration are compared to concentration standards, the WPPO blend performed well.

- The blends of WPPO, ethanol, and EHN have identical temperature characteristics to those of the conventional diesel test fuel especially as the engine speeds hit 75% heading to full engine speed. This was attributed to the presence of ethanol responsible for decreased ignition delay. The presence of high oxygen enrichment was a factor of decreased CO concentration for the tested biodiesels compared with conventional diesel fuel, although there was an increase in CO concentration as fuel CN and blend ratio increased. This is due to deterioration of the combustion characteristics, as the cetane numbers (CNs) and the alcohol blend ratio increase. Nevertheless, this work proposes further study and further investigation on biodiesels with extremely high CN to meet the need for fuel improvers and additives.

Author details

Semakula Maroa[1*] and Freddie Inambao[2]

1 College of Agriculture Science and Engineering, University of KwaZulu-Natal, Durban, South Africa

2 Discipline of Mechanical Engineering, Green Energy Group, Durban, South Africa

*Address all correspondence to: ssemakulamaroa@gmail.com

References

[1] Özgür T, Tosun E, Özgür Tüccar G, Aydin K. Performance, emission and efficiency analysis of a diesel engine operated with diesel and diesel-ethanol (e20) blend. MATTER: International Journal of Science and Technology. 2017;**3**:51-61. DOI: 10.20319/mijst.2017.33.5161

[2] Geng P, Cao E, Tan Q, Wei L. Effects of alternative fuels on the combustion characteristics and emission products from diesel engines: A review. Renewable and Sustainable Energy Reviews. 2017;**71**:523-534. DOI: 10.1016/j.rser.2016.12.080

[3] La Rocca HA, A. Shayler P. Towards keeping diesel fuel supply and demand in balance: Dual-fueling of diesel engines with natural gas. Renewable and Sustainable Energy Reviews. 2017;**70**:666-697. DOI: 10.1016/j.rser.2016.11.249

[4] Asad U, Zheng M. Exhaust gas recirculation for advanced diesel combustion cycles. Applied Energy. 2014;**123**:242-252. DOI: 10.1016/j.apenergy.2014.02.073

[5] Maroa S, Inambao F. The effects of exhaust gas recirculation on the performance and emission characteristics of a diesel engine—A critical review. International Journal of Applied Engineering Research. 2017;**12**:13677-13689

[6] Maroa S, Inambao F. Transportation, pollution and the environment. International Journal of Applied Engineering Research. 2018;**13**:3187-3199

[7] Organization for Economic Co-operation and Development and the International Energy Agency. CO_2 Emissions from Fuel Combustion 2012. Organization for Economic Co-operation and Development. 2013. Available from: http://large.stanford.edu/courses/2013/ph240/rajavi2/docs/co2highlights.pdf [Accessed: 12 October 2019]

[8] Olivier JG, Peters JA, Janssens-Maenhout G. Trends in Global CO_2 Emissions; 2012 Report. 2012. Available from http://large.stanford.edu/courses/2013/ph240/rajavi2/docs/co2highlights.pdf [Accessed: 13 October 2019]

[9] Reşitoğlu İA, Altinişik K, Keskin A. The pollutant emissions from diesel-engine vehicles and exhaust aftertreatment systems. Clean Technologies and Environmental Policy. 2015;**17**:15-27

[10] Song X, Johnson JH, Naber JD. A review of the literature of selective catalytic reduction catalysts integrated into diesel particulate filters. International Journal of Engine Research. 2015;**16**:738-749. DOI: 10.1007/s10098-014-0793-9

[11] Heywood JB. Internal Combustion Engine Fundamentals. New York: McGraw-Hill; 2012

[12] Stone R. Introduction to Internal Combustion Engines. London: Palgrave Macmillan; 2012

[13] Peckham MS, Finch A, Campbell B, Price P, Davies MT. Study of particle number emissions from a turbocharged gasoline direct injection (GDI) engine including data from a fast-response particle size spectrometer. SAE Technical Paper 0148-7191. 2011. DOI: 10.4271/2011-01-1224

[14] World Health Organization. Global Health Risks: Mortality and Burden of Disease Attributable to Selected Major Risks. Geneva: World Health Organization; 2009

[15] Environmental Protection Agency (USA). Potential Revisions to the

Nonroad Engine Regulation Citations in the NSPS for Stationary Internal Combustion Engines, June 28, 2017. 2017. Available from: https://www.epa.gov/stationary-sources-air-pollution/potential-revisions-stationary-internal-combustion-engines-nsps [Accessed: 25 October 2019]

[16] Waluś KJ, Warguła Ł, Krawiec P, Adamiec JM. Legal regulations of restrictions of air pollution made by non-road mobile machinery—The case study for Europe: A review. Environmental Science and Pollution Research. 2018;**25**:3243-3259. DOI: 10.1007/s11356-017-0847-8

[17] Azad A, Rasul M, Khan M, Sharma SC, Bhuiya M. Recent development of biodiesel combustion strategies and modelling for compression ignition engines. Renewable and Sustainable Energy Reviews. 2016;**56**:1068-1086. DOI: 10.1016/j.rser.2015.12.024

[18] Agarwal AK, Agarwal RA, Gupta T, Gurjar BR. Biofuels: Technology, Challenges and Prospects. Berlin: Springer; 2017

[19] Gan S, Ng HK, Pang KM. Homogeneous charge compression ignition (HCCI) combustion: Implementation and effects on pollutants in direct injection diesel engines. Applied Energy. 2011;**88**:559-567. DOI: 10.1016/j.apenergy.2010.09.005

[20] Bendu H, Murugan S. Homogeneous charge compression ignition (HCCI) combustion: Mixture preparation and control strategies in diesel engines. Renewable and Sustainable Energy Reviews. 2014;**38**:732-746. DOI: 10.1016/j.energy.2014.05.086

[21] Zhao Y, Wang Y, Li D, Lei X, Liu S. Combustion and emission characteristics of a DME (dimethyl ether)-diesel dual fuel premixed charge compression ignition engine with EGR (exhaust gas recirculation). Energy. 2014;**72**:608-617. DOI: 10.1016/j.energy.2014.05.086

[22] Jain A, Singh AP, Agarwal AK. Effect of fuel injection parameters on combustion stability and emissions of a mineral diesel fueled partially premixed charge compression ignition (PCCI) engine. Applied Energy. 2017;**190**:658-669. DOI: 10.1016/j.apenergy.2016.12.164

[23] Qiu L, Cheng X, Liu B, Dong S, Bao Z. Partially premixed combustion based on different injection strategies in a light-duty diesel engine. Energy. 2016;**96**:155-165. DOI: 10.1016/j.energy.2015.12.052

[24] Nemati A, Khalilarya S, Jafarmadar S, Khatamnezhad H, Fathi V. Effect of injection characteristics on emissions and combustion of a gasoline fuelled partially-premixed compression ignition engine. International Journal of Automotive Engineering. 2011;**1**:29-37

[25] Yang B, Wang H, Yao M, Zheng Z, Liu J, Ma N, et al. Experimental investigation on the effects of injection strategy on combustion and emission in a heavy-duty diesel engine fueled with gasoline. SAE Technical Paper 0148-7191. 2017. DOI: 10.4271/2017-01-2266

[26] Yousefi A, Guo H, Birouk M. An experimental and numerical study on diesel injection split of a natural gas/diesel dual-fuel engine at a low engine speed. Fuel. 2018;**212**:332-346. DOI: 10.1016/j.fuel.2017.10.053

[27] Saraei SH, Khalilarya S, Jafarmadar S. Numerically modeling of diesel engine and analysis the effects of double injection strategies on performance and pollutant emissions. International Journal of Automotive Engineering. 2016;**6**:2182-2192

[28] Liu H, Ma G, Hu B, Zheng Z, Yao M. Effects of port injection of hydrous

ethanol on combustion and emission characteristics in dual-fuel reactivity-controlled compression ignition (RCCI) mode. Energy. 2017;**145**:592-602. DOI: 10.1016/j.energy.2017.12.089

[29] Zhiqiang Z, Fuquan Z, Jun D, Liguang L, Yuan S. Simulation and optimization of low temperature combustion in diesel engine at high load. Automotive Engineering. 2012;**34**:763-770

[30] Shi L, Xiao W, Li M, Lou L, Deng K-Y. Research on the effects of injection strategy on LTC combustion based on two-stage fuel injection. Energy. 2017;**121**:21-31. DOI: 10.1016/j.energy.2016.12.128

[31] Ramadhas AS, Xu Liu HD, Tian J. Reducing cold start emissions from automotive diesel engine at cold ambient temperatures. Aerosol and Air Quality Research. 2016;**16**:3330-3337. DOI: 10.4209/aaqr.2015.11.0616

[32] Weiss M, Bonnel P, Kühlwein J, Provenza A, Lambrecht U, Alessandrini S, et al. Will Euro 6 reduce the NOx emissions of new diesel cars?—Insights from on-road tests with portable emissions measurement systems (PEMS). Atmospheric Environment. 2012;**62**:657-665. DOI: 10.1016/j.atmosenv.2012.08.056

[33] Johnson BT. Diesel engine emissions and their control. Platinum Metals Review. 2008;**52**:23-37. DOI: 10.1595/147106708x248750

[34] Stanton DW, Lippert AM, Reitz RD, Rutland C. Influence of spray-wall interaction and fuel films on cold starting in direct injection diesel engines. SAE Technical Paper 0148-7191. 1998. DOI: 10.4271/982584

[35] Park Y, Hwang J, Bae C, Kim K, Lee J, Pyo S. Effects of diesel fuel temperature on fuel flow and spray characteristics. Fuel. 2015;**162**:1-7. DOI: 10.1016/j.fuel.2015.09.008

[36] Tian J, Xu H, Sakunthalai RA, Liu D, Tan C, Ghafourian A. Low ambient temperature effects on a modern turbocharged diesel engine running in a driving cycle. SAE International Journal of Fuels and Lubricants. 2014;**7**:726-736. DOI: 10.4271/2014-01-2713

[37] Starck L, Faraj A, Perrin H, Forti L, Jeuland N, Walter B. Cold start on diesel engines: Effect of fuel characteristics. SAE International Journal of Fuels and Lubricants. 2010;**3**:165-174. DOI: 10.4271/2010-01-1506

[38] Liu D, Xu H, Tian J, Tan C, Li Y. Cold and warm start characteristics using HVO and RME blends in a V6 diesel engine. SAE International Journal of Fuels and Lubricants. 2013;**6**:478-485. DOI: 10.4271/2013-01-1306

[39] Brown N, Gupta V, Rocca AL, Shayler P, Murphy M, Pegg I, et al. Investigations of fuel injection strategy for cold starting direct-injection diesel engines. Proceedings of the Institution of Mechanical Engineers, Part D: Journal of Automobile Engineering. 2007;**221**:1415-1424. DOI: 10.1243/09544070JAUTO473

[40] Thom R. Emmission–Immission Regularien Eine Aundauernde Herausforderung. In: 3rd Conference on Emission Control, Dresden, Germany. May 2006

[41] Steinparzen F. Dieselmotoren für PKW: Gestern–Heute–Morgen. In: 18th AVL Conference, Engine and Environment, Helmut-List-Halle Conference Centre, Graz, Austria, 7-8 September, 2006

[42] Cooper B, Penny I, Beasley M, Greaney A, Crump J. Advanced diesel technology to achieve tier 2 bin 5 emissions compliance in US light-duty diesel applications; SAE Technical Paper 0148-7191. 2006. DOI: 10.4271/2006-01-1145

[43] Park S-K, Lee J, Kim H-M. Experimental study on the spiral and oval spiral EGR cooler efficiencies in a diesel engine. Heat and Mass Transfer. 2014;**50**:1783-1789. DOI: 10.1007/s00231-014-1362-9

[44] Adomeit P, Pischinger S, Becker M, Rohs H, Greis A. Laser optical diagnostics and numerical analysis of HSDI combustion systems. In: Proceedings of Conference on Thermo- and Fluid Dynamic Processes in Diesel Engines 2. Thiesel. Valencia, Spain; 11-13 September 2002. p. 331

[45] Pickett L. Soot formation at low flame temperature diesel operating conditions. In: 9th International Conference, Present and Future Engines for Automobiles. San Antonio, TX, USA: SAE Transactions; 2005

[46] Nova I, Tronconi E. Urea-SCR Technology for deNO_x after Treatment of Diesel Exhausts. Berlin: Springer; 2014

[47] Twigg MV. Urea-SCR technology for deNO_x after treatment of diesel exhausts. Book review. Johnson Matthey Technology Review. 2015;**59**:221-232

[48] Walker A. Future challenges and incoming solutions in emission control for heavy-duty diesel vehicles. Topics in Catalysis. 2016;**59**:695-707. DOI: 10.1007/s11244-016-0540-z

[49] Thangaraja J, Kannan C. Effect of exhaust gas recirculation on advanced diesel combustion and alternate fuels—A review. Applied Energy. 2016;**180**:169-184. DOI: 10.1016/j.apenergy.2016.07.096

[50] Divekar PS, Chen X, Tjong J, Zheng M. Energy efficiency impact of EGR on organizing clean combustion in diesel engines. Energy Conversion and Management. 2016;**112**:369-381. DOI: 10.1016/j.enconman.2016.01.042

[51] Schnitzler J. Particulate Matter and NO_x Exhaust after Treatment Systems. Aachen, Germany: FEV Metroethnic GmbH; 2006

[52] Ward C. Diesel exhaust after-treatment maintenance. Tech Directions. 2017;**76**:26

[53] Yun BK, Kim MY. Modeling the selective catalytic reduction of NO_x by ammonia over a Vanadia-based catalyst from heavy duty diesel exhaust gases. Applied Thermal Engineering. 2013;**50**:152-158. DOI: 10.1016/j.applthermaleng.2012.05.039

[54] Azama A, Alia S, Iqbala A. Emissions from diesel engine and exhaust after treatment technologies. In: 4th International Conference on Energy, Environment and Sustainable Development 2016 (EESD 2016); 2016

[55] Turns SR. An Introduction to Combustion. Vol. 499. New York: McGraw-Hill; 1996

[56] Zhang Y, Lou D, Tan P, Hu Z, Feng Q. Experimental study on particulate emission characteristics of an urban bus equipped with CCRT after-treatment system fuelled with biodiesel blend. SAE Technical Paper 0148-7191. 2017. DOI: 10.4271/2017-01-0933

[57] Guardiola C, Pla B, Piqueras P, Mora J, Lefebvre D. Model-based passive and active diagnostics strategies for diesel oxidation catalysts. Applied Thermal Engineering. 2017;**110**:962-971. DOI: 10.1016/j.applthermaleng.2016.08.207

[58] Chen K-S, Lin Y-C, Hsu K-H, Wang H-K. Improving biodiesel yields from waste cooking oil by using sodium methoxide and a microwave heating system. Energy. 2012;**38**:151-156. DOI: 10.1016/j.energy.2011.12.020

[59] American Petroleum Institute. HPV Testing Group Kerosene/Jet Fuel Category Assessment

Document. Submitted to the US EPA by the American Petroleum Institute, Consortium Registration 1100997; 2010

[60] Harley RA, Kean AJ. Chemical Composition of Vehicle-Related Volatile Organic Compound Emission in Central California. USA: Department of Civil and Environmental Engineering, University of California; 2004

[61] Ghosh P, Hickey KJ, Jaffe SB. Development of a detailed gasoline composition-based octane model. Industrial and Engineering Chemistry Research. 2006;**45**:337-345. DOI: 10.1021/ie050811h

[62] Lapuerta M, Armas O, Hernández JJ, Tsolakis A. Potential for reducing emissions in a diesel engine by fuelling with conventional biodiesel and Fischer–Tropsch diesel. Fuel. 2010;**89**:3106-3113. DOI: 10.1016/j.fuel.2010.05.013

[63] Bhaskar K, Nagarajan G, Sampath S. The performance and emission characteristics of fish oil methyl esters (fome) and diesel blends in a partially premixed charge compression ignition engine. International Journal of Green Energy. 2014;**11**:389-403. DOI: 10.1080/15435075.2013.773514

[64] Venkatesan H, Sivamani S, Bhutoria K, Vora HH. Assessment of waste plastic oil blends on performance, combustion and emission parameters in direct injection compression ignition engine. International Journal of Ambient Energy. 2017;**40**:1-9. DOI: 10.1080/01430750.2017.1381155

[65] Kaimal VK, Vijayabalan P. An investigation on the effects of using DEE additive in a DI diesel engine fuelled with waste plastic oil. Fuel. 2016;**180**:90-96. DOI: 10.1016/j.fuel.2016.04.030

[66] Karjalainen P, Ntziachristos L, Murtonen T, Wihersaari H, Simonen P, Mylläri F, et al. Heavy-duty diesel exhaust particles during engine motoring formed by lube oil consumption. Environmental Science & Technology. 2016;**50**:12504-12511. DOI: 10.1021/acs.est.6b03284

[67] Lee D, Choi S, Lee C. Impact of SME blended fuel combustion on soot morphological characteristics in a diesel engine. International Journal of Automotive Technology. 2013;**14**:757-762. DOI: 10.1007/s12239-013-0083-2

[68] Ghobadian B, Rahimi H, Nikbakht A, Najafi G, Yusaf T. Diesel engine performance and exhaust emission analysis using waste cooking biodiesel fuel with an artificial neural network. Renewable Energy. 2009;**34**:976-982. DOI: 10.1016/j.renene.2008.08.008

[69] Ganesh D, Gowrishankar G. Effect of nano-fuel additive on emission reduction in a biodiesel fuelled CI engine. International Conference on Electrical and Control Engineering. 2011;**2011**:3453-3459. DOI: 10.1109/ICECENG.2011.6058240

[70] Babu P, Rao K, Rao B, Srivastava S, Beohar H, Gupta B. The role of oxygenated fuel additive (DEE) along with Mahuva methyl ester to estimate performance and emission analysis of DI-diesel engine. International Journal of Thermal Technologies. 2012;**2**:119-123

[71] Fattah IR, Masjuki H, Kalam M, Wakil M, Ashraful A, Shahir SA. Experimental investigation of performance and regulated emissions of a diesel engine with *Calophyllum inophyllum* biodiesel blends accompanied by oxidation inhibitors. Energy Conversion and Management. 2014;**83**:232-240. DOI: 10.1016/j.enconman.2014.03.069

[72] Kalam M, Masjuki H. Testing palm biodiesel and NPAA additives to

control NO_x and CO while improving efficiency in diesel engines. Biomass and Bioenergy. 2008;**32**:1116-1122. DOI: 10.1016/j.biombioe.2008.02.009

[73] Tutak W, Lukács K, Szwaja S, Bereczky Á. Alcohol-diesel fuel combustion in the compression ignition engine. Fuel. 2015;**154**:196-206. DOI: 10.1016/j.fuel.2015.03.071

[74] Lujaji F, Kristóf L, Bereczky A, Mbarawa M. Experimental investigation of fuel properties, engine performance, combustion and emissions of blends containing croton oil, butanol, and diesel on a CI engine. Fuel. 2011;**90**: 505-510. DOI: 10.1016/j.fuel.2010.10.004

[75] Sanli H, Canakci M, Alptekin E, Turkcan A, Ozsezen A. Effects of waste frying oil-based methyl and ethyl ester biodiesel fuels on the performance, combustion and emission characteristics of a DI diesel engine. Fuel. 2015;**159**:179-187. DOI: 10.1016/j.fuel.2015.06.081

[76] Shirneshan A. HC, CO, CO_2 and NO_x emission evaluation of a diesel engine fueled with waste frying oil methyl ester. Procedia—Social and Behavioral Sciences. 2013;**75**:292-297. DOI: 10.1016/j.sbspro.2013.04.033

[77] Environmental Protection Agency. Inventory of US Greenhouse Gas Emissions and Sinks: 1990-2009. Washington, D.C., United States: Environmental Protection Agency; 2012

[78] Rahman MM, Hassan MH, Kalam MA, Atabani AE, Memon LA, Rahman SA. Performance and emission analysis of *Jatropha curcas* and *Moringa oleifera* methyl ester fuel blends in a multi-cylinder diesel engine. Journal of Cleaner Production. 2014;**65**:304-310. DOI: 10.1016/j.jclepro.2013.08.034

[79] İçıngür Y, Altiparmak D. Effect of fuel cetane number and injection pressure on a DI diesel engine performance and emissions. Energy Conversion and Management. 2003;**44**:389-397. DOI: 10.1016/S0196-8904(02)00063-8

[80] Wu F, Wang J, Chen W, Shuai S. A study on emission performance of a diesel engine fueled with five typical methyl ester biodiesels. Atmospheric Environment. 2009;**43**:1481-1485. DOI: 10.1016/j.atmosenv.2008.12.007

[81] Patil A, Taji S, Relekar MAN, Patil MAN, Dhananjay M, Khamkar M, et al. Investigation on effect of oxygenated additive on multicylinder diesel engine performance and emission. International Journal of Innovative Research in Science, Engineering and Technology. 2014;**3**:14113-14120

[82] Lin SL, Lee WJ, Lee CF, Chen SJ. Energy savings and emission reduction of nitrogen oxides, particulate matter, and polycyclic aromatic hydrocarbons by adding water-containing acetone and neat soybean oil to a diesel-fueled engine generator. Energy & Fuels. 2010;**24**:4522-4533. DOI: 10.1021/ef100556b

[83] Wang Y, Liu H, Lee C-FF. Particulate matter emission characteristics of diesel engines with biodiesel or biodiesel blending: A review. Renewable and Sustainable Energy Reviews. 2016;**64**:569-581. DOI: 10.1016/j.rser.2016.06.062

[84] Apte JS, Marshall JD, Cohen AJ, Brauer M. Addressing global mortality from ambient PM2.5. Environmental Science & Technology. 2015;**49**:8057-8066. DOI: 10.1021/acs.est.5b01236

[85] Barrios CC, Domínguez-Sáez A, Martín C, Álvarez P. Effects of animal fat-based biodiesel on a TDI diesel engine performance, combustion characteristics and particle number and size distribution emissions. Fuel.

2014;**117**:618-623. DOI: 10.1016/j.fuel.2013.09.037

[86] Agrawal AK, Singh SK, Sinha S, Shukla MK. Effect of EGR on the exhaust gas temperature and exhaust opacity in compression ignition engines. Sadhana. 2004;**29**:275-284. DOI: 10.1007/BF02703777

[87] Thiruvengadam A, Besch MC, Yoon S, Collins J, Kappanna H, Carder DK, et al. Characterization of particulate matter emissions from a current technology natural gas engine. Environmental Science & Technology. 2014;**48**:8235-8242. DOI: 10.1021/es5005973

[88] Kittelson D, Kraft M. Particle formation and models. In: Encyclopedia of Automotive Engineering. Hoboken: Wiley Publishing Company; Available from: https://doi.org/10.1002/9781118354179.auto161

[89] Tighe C, Twigg M, Hayhurst A, Dennis J. The kinetics of oxidation of diesel soots by NO_2. Combustion and Flame. 2012;**159**:77-90. DOI: 10.1021/es903672y

[90] Chuepeng S, Xu H, Tsolakis A, Wyszynski M, Price P, Stone R, et al. Particulate emissions from a common rail fuel injection diesel engine with RME-based biodiesel blended fuelling using thermo-gravimetric analysis. SAE Technical Paper 0148-7191. 2008. DOI: 10.4271/2008-01-0074

[91] Williams A, McCormick RL, Hayes RR, Ireland J, Fang HL. Effect of biodiesel blends on diesel particulate filter performance. SAE Technical Paper 0148-7191. 2006

[92] Cheung KL, Polidori A, Ntziachristos L, Tzamkiozis T, Samaras Z, Cassee FR, et al. Chemical characteristics and oxidative potential of particulate matter emissions from gasoline, diesel, and biodiesel cars. Environmental Science & Technology. 2009;**4**:6334-6340. DOI: 10.1021/es900819t

[93] Agarwal AK, Gupta T, Shukla PC, Dhar A. Particulate emissions from biodiesel fuelled CI engines. Energy Conversion and Management. 2015;**94**:311-330. DOI: 10.1016/j.enconman.2014.12.094

[94] Oprescu E-E, Dragomir RE, Radu E, Radu A, Velea S, Bolocan I, et al. Performance and emission characteristics of diesel engine powered with diesel–glycerol derivatives blends. Fuel Processing Technology. 2014;**126**:460-468. DOI: 10.1016/j.fuproc.2014.05.027

[95] Ashraful A, Masjuki H, Kalam M. Particulate matter, carbon emissions and elemental compositions from a diesel engine exhaust fuelled with diesel–biodiesel blends. Atmospheric Environment. 2015;**120**:463-474. DOI: 10.1016/j.atmosenv.2015.09.028

[96] Cheng M-T, Chen H-J, Young L-H, Yang H-H, Tsai YI, Wang L-C, et al. Carbonaceous composition changes of heavy-duty diesel engine particles in relation to biodiesels, after-treatments and engine speeds. Journal of Hazardous Materials. 2015;**297**:234-240. DOI: 10.1016/j.jhazmat.2015.04.076

[97] Dorenkamp R. Application of a New Filter Material in Volkswagen's Diesel Particulate Filter System. In: 3rd Conference on Emission Control, Dresden, Germany; May 2006

[98] Boretto G, Imarisio R, Rellecati P, Barucchi E, Sanguedolce A. Serial Application of a Catalyzed Particulate Filter on Common Rail DI Diesel Engines for Passenger Cars. Barcelona, Spain: F2004V068, FISITA World Automotive Congress; 2004. pp. 23-27

[99] Rothe D, Knauer M, Emmerling G, Deyerling D, Niessner R. Emissions during active regeneration of a diesel particulate filter on a heavy-duty diesel

engine: Stationary tests. Journal of Aerosol Science. 2015;**90**:14-25. DOI: 10.1016/j.jaerosci.2015.07.007

[100] Goersmann C. PM control systems with low NO_2 emissions. In: 3rd Conference on Emission Control, Dresden, Germany; May 2006

[101] Zink UH, Johnson TV. State-of-the-art filter regeneration management—Concepts realized by LDV companies. In: U.S. Dept. of Energy Diesel Engine Emissions Reduction (DEER) Conference; 21-25 August 2005; Chicago, IL, U.S.A. 2005

[102] Lenz HP. 38th international—Vienna motor symposium. Sonderprojekte ATZ/MTZ. 2017;**22**:34-61

[103] Agarwal A, Gupta KT, Kothari A. Particulate emissions from biodiesel vs diesel fuelled compression ignition engine. Renewable and Sustainable Energy Reviews. 2011;**15**:3278-3300. DOI: 10.1016/j.rser.2011.04.002

[104] Zhu L, Cheung C, Huang Z. A comparison of particulate emission for rapeseed oil methyl ester, palm oil methyl ester and soybean oil methyl ester in perspective of their fatty ester composition. Applied Thermal Engineering. 2016;**94**:249-255. DOI: 10.1016/j.applthermaleng.2015.10.132

[105] Di Iorio S, Magno A, Mancaruso E, Vaglieco B, Arnone L, Dal Bello L. Engine performance and emissions of a small diesel engine fueled with various diesel/RME blends. SAE Paper: 32-0135. 2014. DOI: 10.4271/2014-32-0135

[106] Turns SR. An Introduction to Combustion. Vol. 287. New Delhi: McGraw-hill Education India; 2012

[107] Hoekman SK, Robbins C. Review of the effects of biodiesel on NO_x emissions. Fuel Processing Technology. 2012;**96**:237-249. DOI: 10.1016/j.fuproc.2011.12.036

[108] Kim K-H, Kabir E, Kabir S. A review on the human health impact of airborne particulate matter. Environment International. 2015;**74**:136-143. DOI: 10.1016/j.envint.2014.10.005

[109] Nakatsuji T, Matsubara M, Rouistenmäki J, Sato N, Ohno H. A NO_x reduction system using ammonia-storage selective catalytic reduction in rich/lean excursions. Applied Catalysis B: Environmental. 2007;**77**:190-201. DOI: 10.1016/j.apcatb.2007.07.018

[110] Marchitti F, Nova Hemings EB, Forzatti IP, Tronconi E. Enhancing the low-T NH_3-SCR activity of a commercial Fe-zeolite catalyst by NH_4NO_3 dosing: An experimental and modeling study. Emission Control Science and Technology. 2016;**2**:1-9. DOI: 10.1007/s40825-015-0030-y

[111] Myer AM, Holl SM, Tendolkar O, Crofts JD, Johnson S, Khot AD. Ammonia generation system for an SCR system. Google Patents; 2016

[112] Kowatari T, Hamada Y, Amou K, Hamada I, Funabashi H, Takakura T. et al. A study of a new aftertreatment system (1): A new dosing device for enhancing low temperature performance of urea-SCR. SAE Technical Paper 0148-7191. 2006

[113] Johnson TV. Review of diesel emissions and control. SAE International Journal of Fuels and Lubricants. 2010;**3**:16-29. DOI: 10.1243/14680874JER04009

[114] Johnson TV. Review of vehicular emissions trends. SAE International Journal of Engines. 2015;**8**:1152-1167. DOI: 10.4271/2015-01-0993

[115] Theis JR, Ura JA, Goralski CT, Caine J, Davies M, Kay D, et al. The effects of aging temperature and PGM loading on the NO_x storage capacity of a lean NO_x trap. SAE

Technical Paper 0148-7191. 2005. DOI: 10.4271/2005-01-1117

[116] Leet JA, Simescu S, Froelund K, Dodge LG, Roberts CE. Emissions solutions for 2007 and 2010 heavy-duty diesel engines. SAE Technical Paper 0148-7191. 2004. DOI: 10.4271/2004-01-0124

[117] Chilumukuru K, Gupta A, Ruth M, Cunningham M, Kothandaraman G, Cumaranatunge L, et al. Aftertreatment architecture and control methodologies for future light duty diesel emission regulations. SAE International Journal of Engines. 2017;**10**:1580-1587. DOI: 10.4271/2017-01-0911

[118] Guan B, Zhan R, Lin H, Huang Z. Review of state-of-the-art technologies of selective catalytic reduction of NO_x from diesel engine exhaust. Applied Thermal Engineering. 2014;**66**:395-414. DOI: 10.1016/j.applthermaleng.2014.02.021

[119] Wittka T, Holderbaum B, Dittmann P, Pischinger S. Experimental investigation of combined LNT+ SCR diesel exhaust aftertreatment. Emission Control Science and Technology. 2015;**1**:167-182. DOI: 10.1007/s40825-015-0012-0

[120] Ming-Cheng W, Han T, Fisher GB. Experimental evaluation of reformate-assisted diesel NO_x trap desulfation. SAE Technical Paper 0148-7191. 2005. DOI: 10.4271/2005-01-3878

[121] Rao KS. Effect of EGR on diesel engine performance and exhaust emission running with cotton seed biodiesel. International Journal of Mechanical and Mechatronic Engineering. 2016;**16**:64-69

[122] Bowman CT. Control of combustion-generated nitrogen oxide emissions: Technology driven by regulation. In Symposium (International) on Combustion. 1992:859-878

[123] Heywood JB. Internal Combustion Engine Fundamentals Vol. 930. New York: McGraw-Hill; 1988

[124] Benjumea P, Agudelo JR, Agudelo AF. Effect of the degree of unsaturation of biodiesel fuels on engine performance, combustion characteristics, and emissions. Energy & Fuels. 2010;**25**:77-85. DOI: 10.1021/ef101096x

[125] Altun Ş. Effect of the degree of unsaturation of biodiesel fuels on the exhaust emissions of a diesel power generator. Fuel. 2014;**11**:450-457. DOI: 10.1016/j.fuel.2013.09.028

[126] Zhu Z, Guo H, Zhou A, Li D, Liu S. One way to reduce the NO_x emission of biodiesels: The increase of cetane number. International Journal of Green Energy. 2016;**13**:957-962. DOI: 10.1080/15435075.2011.647366

Chapter 4

Investigation of the Gasoline Engine Performance and Emissions Working on Methanol-Gasoline Blends Using Engine Simulation

Simeon Iliev

Abstract

The aim of this study is to develop the one-dimensional model of a four-cylinder, four-stroke, multi-point injection system SI engine and a direct injection system SI engine for predicting the effect of various fuel types on engine performances, specific fuel consumption, and emissions. Commercial software AVL BOOST was used to examine the engine characteristics for different blends of methanol and gasoline (by volume: 5% methanol [M5], 10% methanol [M10], 20% methanol [M20], 30% methanol [M30], and 50% methanol [M50]). The methanol-gasoline fuel blend results were compared to those of net gasoline fuel. The obtained results show that when methanol-gasoline fuel blends were used, engine performance such as power and torque increases and the brake-specific fuel consumption increases with increasing methanol percentage in the blended fuel.

Keywords: methanol blends, alternative fuels, spark-ignition engine, emissions, engine simulation

1. Introduction

Alternative fuels are derived from resources other than petroleum. When using these fuels in internal combustion engines (ICE), they produce less air pollution emissions than gasoline. Most of them are more economically beneficial than fossil fuels. Last but not least, they are renewable. The most commonly used alternative fuels are natural gas, propane, methanol, ethanol, and hydrogen. Lots of works have been written on engine operating with these fuels individually, but very few compared some of these alternative fuels together in the same engine [1–3]. The idea of adding low contents of ethanol or methanol to gasoline is not new, extending back at least to the 1970s, when oil supplies were reduced and a search for alternative energy carriers began in order to replace gasoline and diesel fuel. Initially, methanol was considered the most attractive alcohol to be added to gasoline. Methanol production can be from biomass, coal, or natural gas with acceptable energy costs. The gasification of biomass can lead to methanol, mixed alcohols, and Fischer-Tropsch liquids [4]. Since methanol can be produced from natural gas at no great cost, and is quite easy to blend with gasoline, this alcohol was seen as an attractive additive. Methanol can also be used in pure form in internal combustion engines; the fact that it is a liquid fuel makes it suitable for storing and distributing. It does produce hydrocarbon emissions similar to gasoline (different species); its single-carbon-molecule nature and combustion characteristics mean that its emissions of oxides of nitrogen and particulate matter are significantly lower than hydrocarbon fuels. However, when using methanol in practice, it became clear that precautions had to be taken when handling it and that methanol is aggressive to some materials, such as plastic components and even metals in the fuel system [5].

Methanol has many advantages (characteristics) that make it very suitable for use as a fuel in spark-ignition engine. Some of these characteristics are given in a **Table 1**, and they are as follows:

- High molar expansion ratio
- High hydrogen-to-carbon ratio
- Being liquid at standard temperature and pressure
- High heat of vaporization ("latent heat")
- High flame speed
- Low combustion temperature
- High specific energy ratio (i.e., energy per unit of fuel-air mixture)

Methanol is the simplest alcohol and is usually referred to as the "light" alcohol. It is the simplest carbonaceous molecule that is liquid at standard temperature and pressure. This makes it easy to store and transport with minimal losses on the vehicle and in the fuel infrastructure. It is also known as methyl alcohol. The higher autoignition temperature of methanol compared to gasoline allows the engines to operate at a higher compression ratio; thereby they can be more efficient.

The methanol consists of just one molecule unlike gasoline, diesel, kerosene, etc. which the properties can change depending on the source, and as such it is easier to simulate the process for. The molecular weight of methanol is approximately four times lighter than gasoline. The diffusion rate of lighter fuel is lower than that of heavier fuel, and it results in lower emission.

Adding methanol to gasoline allows the fuel mixture to combust more completely due to the presence of oxygen (inherent oxygen in its molecular structure), which increases the combustion efficiency and reduces the emission of CO and NOx by converting them into CO_2 and NO_2. Besides, methanol does not contain sulfur or complex organic compounds [9], resulting in zero emission of sulfur-based pollutants (SO_2 and SO_3, which are responsible for acid rain). The organic emissions (ozone precursors) from alcohol combustion have lower reactivity, which can stimulate ozone formation [10].

Methanol has a higher latent heat of vaporization than gasoline (**Table 1**). It provides a cooling effect on the intake charge compared to gasoline. This effect improves the brake thermal efficiency and power output. The lower caloric value of methanol due to oxygen content in its molecular structure requires higher fuel quantity to be injected in order to achieve an equivalent brake power output.

Properties	Gasoline	Methanol
Chemical formula	C_8H_{15}	CH_3OH
Molar mass, kg/kmol	114	32
Oxygen content, wt%	—	50
Carbon content, wt%	86	38
Hydrogen content, wt%	14	12
Stoichiometric AFR	14,5	6,43
Lower heating value, MJ/kg	44,3	20,1
Higher heating value, MJ/kg	48	22,8
Volumetric energy content, MJ/m^3	31,746	15,871
Heat of evaporation, kJ/kg at 1 bar	375	1089
Research octane number	96,5	112
Motor octane number	87,2	91
Cetane number	—	<5
Boiling temperature, °C at 1 bar	27–245	65
Vapor pressure, bar at 20°C	0,25–0,45	0,13
Critical pressure, bar	—	81
Critical temperature, °C	—	239,4
Kinematic viscosity, cSt at 20°C	0,6	0,74
Destiny, kg/cm^3	740	798
Surface tension, N/m at 20°C	—	0,023
Minimum ignition energy, mJ at $\varphi = 1$	0,8	0,21
Autoignition temperature, °C	246–280	470
Peak flame temperature, °C at 1 bar	2030	1890
Adiabatic flame temperature, K	~ 2275	2143
Flammability limits (vol%)	1,4–7,6	6–36
Flash point, °C	−45	12
Bulk modulus, N/mm^2 at 20°C 2 MPa	1300	823
Specific CO_2 emissions, g/MJ	73,95	68,44
Specific CO_2 emissions relative to gasoline	1	0,93

Table 1.
Comparison of fuel properties [6–8].

The major issue encountered when blending water, methanol, and gasoline is phase separation. The critical phase separation temperature of methanol-gasoline blends increases with the amount of water present in the blend. Because of this very small water tolerance of the methanol-gasoline blend, water contamination during methanol transport and storage has to be avoided [11]. The blends with gasoline and low methanol concentrations will increase the vapor pressure.

Another important problem is related to the engine cold starting of very high blend alcohols in gasoline. Because of lower energy density and higher heat of vaporization of methanol, more mass needs to evaporate and therefore more energy. The lower flammability limit of methanol is higher than that of gasoline which is also the reason for cold starting [12].

Methanol has a higher octane number than pure gasoline fuel [13]. This enables higher compression ratios of engines and, as a result, increases its thermal efficiency [14].

There are many publications with different blends of alcohols and gasoline fuel. For example, Shenghua et al. [15] used a gasoline engine to examine different percentages of methanol blends (from 10 to 30%) in gasoline. From the results obtained, it has been established that power and engine torque decreased, whereas the brake thermal efficiency improved with the increase of methanol percentage in the fuel blend. Another study [16] has studied the influence of methanol-gasoline blends on the gasoline engine performance. The results obtained showed that the highest brake mean effective pressure (BMEP) was obtained from 5% methanol-gasoline blend. In another study, Altun et al. [17] studied the influence of methanol and ethanol blending (5 and 10%) in gasoline fuel on engine performance and emissions. Blended fuels showed the best result in emissions. The emissions of HC are reduced by 13 and 15% for E10 and M10. The results obtained show a decrease in CO emissions by 10,6 and 9,8%, but CO_2 emission increased for E10 and M10. The blended fuels with methanol and ethanol showed an increase in the brake-specific fuel consumption and a decrease in break thermal efficiency compared to gasoline. Some authors suggested that the oxygenated nature of alcohols can lead to more complete combustion and consequently to reduced engine-out CO emissions [18, 19]. Liang et al. [20] studied PM emission from gasoline direct-injected engine and port fuel-injected engine fueled by gasoline and methanol-gasoline blend M15. They found that the PM emission was lower for M15 than for gasoline.

2. Research methodology

The aim of the present chapter is to develop the one-dimensional model of four-stroke port fuel injection (PFI) gasoline engine and four-stroke direct injection (GDI) gasoline engine for predicting the effect of methanol-gasoline (M0–M50) addition to gasoline on the exhaust emissions and performance of gasoline engine. For this, simulation of gasoline SI engine (calibrated) as the basic operating condition and the laminar burning velocity correlations of methanol-gasoline blends for calculating the changed combustion duration were used. The engine power, specific fuel consumption, and exhaust emissions were compared and discussed [21, 22].

Computer simulation is becoming an important tool for time and cost efficiency in an engine's development. The simulation results are challenging to be obtained experimentally. Using computational fluid dynamics (CFD) has allowed researchers to understand the flow behavior and quantify important flow parameters such as mass flow rates or pressure drops, under the condition that the CFD tools have been properly validated against experimental results.

CFD software products include KIVA, AVL FIRE, AVSYS, STAR-CD, VECTIS, FLUENT, PHOENICS, Flow Vision, and more. The above programs allow to model with great accuracy the modeling of gases, the movement of the dispersed fuel in the combustion chamber of the engine, the movement of the thin layer of fuel formed on the surface of solid walls, the temperature field, and other phenomena. The fluid-structure interaction analysis, successfully implemented in the Ansys program, integrates state-of-the-art computational tools related to fluid and gas mechanics and solid-state mechanics to allow a multidisciplinary research.

The software for thermodynamic and gas-dynamic calculations include AVL BOOST, Ricardo WAVE, GT-Power, and others. These software products are characterized by a well-developed user interface that includes one-dimensional and multidimensional models.

2.1 Simulation setup

The simulation tools are the most used in recent years owing to its continuous increase in computational power. The use of engine simulations enables optimization of engine combustion, geometry, and operating characteristics toward improving specific fuel consumption and exhaust emissions and reducing engine development time and costs. Consequently, it can be expected that the use of engine simulations during engine construction will continue to increase. Engine modeling is a fruitful research area, and therefore many laboratories have their own engine thermodynamic models with varying degrees of complexity, scope, and ease of use [23].

Many researchers develop their own computer code describing different processes of engine operation. One of the studies [24] developed the computer code for simulating spark-ignited engine using alternative fuels, and results were validated with experimental data. The engine model is a quasi-dimensional two-zone model including ordinary differential equations for describing dynamical behavior during the intake, compression, power, and exhaust strokes. The engine model uses the Woschni correlation to estimate engine heat transfer. Another author [25] created a model for simulating the performance of spark-ignition engines fueled with gasoline and ethanol fuels and their mixtures. In this model the combustion chamber was divided into burned and unburned zones separated by a flame front. The pressure was assumed to be uniform throughout the cylinder charge. The instantaneous heat interaction between the burned and unburned zones and its walls was calculated by using the semiempirical expression for a four-stroke engine [26].

The one-dimensional engine simulation is widely used for design, development, calibration, and optimization because they make it possible for the entire engine to be modeled, they do not require high computing power, and the calculations are performed in a relatively short time [27, 28]. The one-dimensional (1D) engine model consists of sub-models of selected processes that can be investigated using more detailed modeling approaches (quasi-dimensional or three-dimensional models) to increase the accuracy of the overall engine simulation results.

The model of combustion as part of one-dimensional engine simulations provides the burning rate that represents the heat release rate in the combustion process for a given fuel blend, engine geometry, and set of operating conditions. The burning rate can be computed empirically and or derived from physical, detailed coupled turbulent flames, or chemical kinetic correlations of combustion processes.

The one-dimensional model of SI engine is created by the AVL BOOST software and has been employed to examine the emissions and performance working on gasoline and methanol-gasoline blends. The preprocessing steps of AVL BOOST enables the user to build a one-dimensional engine test bench setup using the predefined elements provided in the software toolbox. The various elements are joined by the desired connectors to establish the complete engine model using pipelines. It is important to make a correct estimate of the boundary conditions as it directly affects the accuracy of the results. Having a proper combustion model will enhance the understanding of the physical phenomena, including the effects of valve phasing, type of fuel, compression ratio, exhaust gas recirculation, etc., and, thus, enable comprehensive design and optimization of the engine.

In **Figure 1**, PFIE symbolizes the engine, while C1 to C4 are the cylinders of the SI engine. The cylinders of the engine are the main element in this model, because they have many very important parameters to settle: the bore, stroke, internal geometry, connecting rod, length and compression ratio, piston pin offset, and the mean crankcase pressure. MP1 to MP18 symbolize the measuring points. The plenum is market with PL1 to PL4. System boundary are market with SB1 and SB2. The

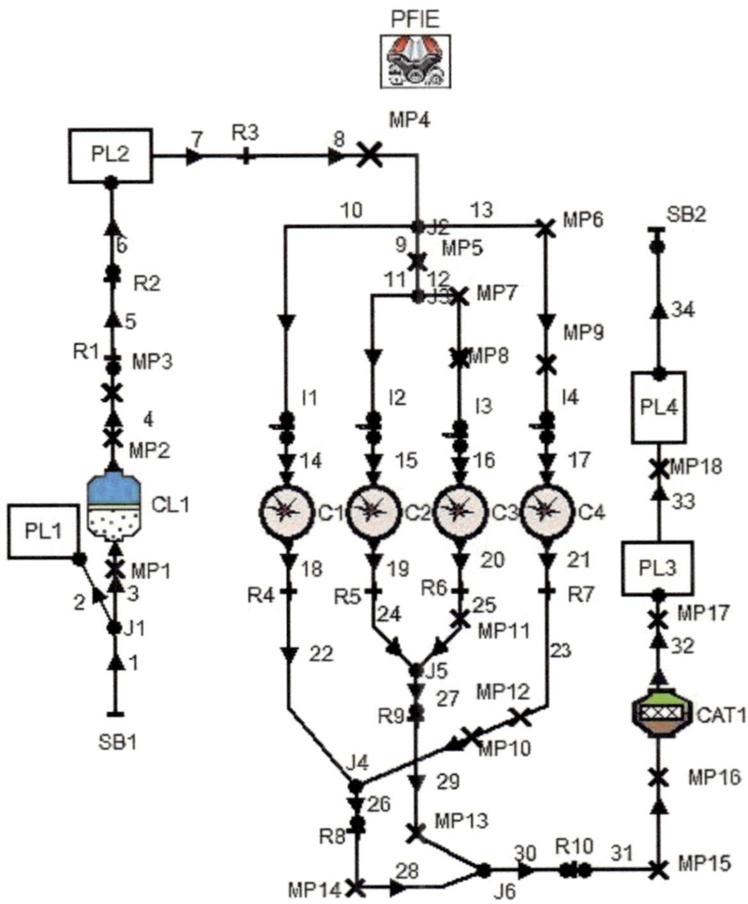

Figure 1.
Schematic of the gasoline PFI engine model.

cleaner is market with CL1. R1 to R10 stand for flow restrictions, CAT1 symbolize catalyst and fuel injectors—I1 to I4 are. The flow pipes are numbered 1 to 34.

The calibrated gasoline PFI engine model was described by Iliev [29], and its layout is shown in **Figure 1** with engine specification shown in **Table 2**.

The schematic of the calibrated GDI engine model is shown in **Figure 2** with engine specification shown in **Table 3**.

The comparison between the properties of gasoline and methanol is presented in **Table 1**. As shown in **Table 1**, compared with gasoline, methanol has a higher elemental oxygen content and a lower heating value, molecular weight, elemental carbon, hydrogen content and stoichiometric air/fuel ratio (AFR).

2.2 Description of combustion model

In this research, the Vibe (two-zone) model was chosen for simulation analysis of the combustion. The combustion chamber was divided into two regions: unburned gas region and burned gas regions [17]. For the burned charge and unburned charge, the first law of thermodynamics is applied:

$$\frac{dm_b u_b}{d\alpha} = -p_c \frac{dV_b}{d\alpha} + \frac{dQ_F}{d\alpha} - \sum \frac{dQ_{Wb}}{d\alpha} + h_u \frac{dm_b}{d\alpha} - h_{BB,b} \frac{dm_{BB,b}}{d\alpha} \quad (1)$$

Engine parameters	Value
Bore	86 (mm)
Stroke	86 (mm)
Compression ratio	10.5
Connection rod length	143.5 (mm)
Number of cylinders	4
Piston pin offset	0 (mm)
Displacement	2000 (cc)
Intake valve open	20 BTDC (deg)
Intake valve close	70 ABDC (deg)
Exhaust valve open	50 BBDC (deg)
Exhaust valve close	30 ATDC (deg)
Piston surface area	5809 (mm^2)
Cylinder surface area	7550 (mm^2)
Number of strokes	4

Table 2.
PFI engine specification.

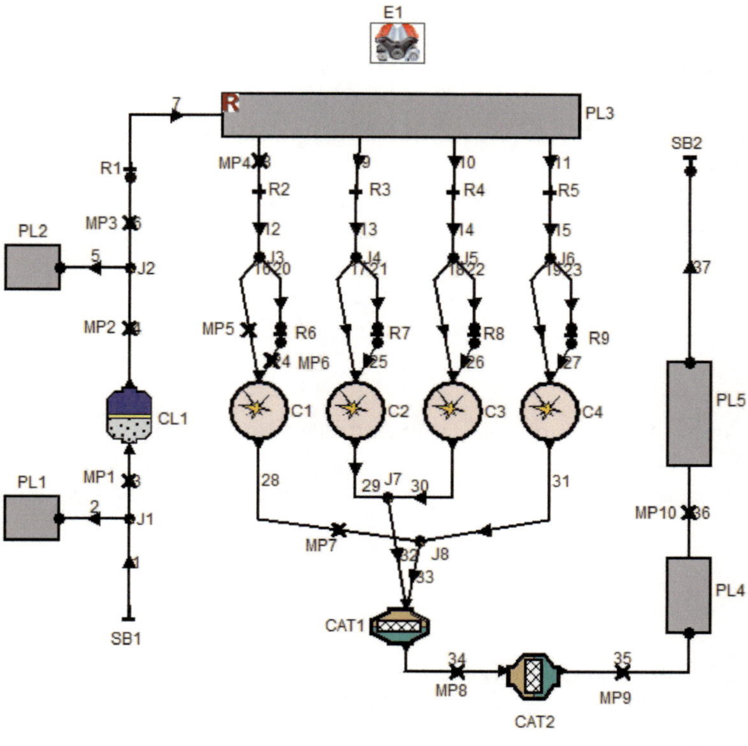

Figure 2.
Schematic of the gasoline GDI engine model.

Engine parameters	Value
Bore	80 (mm)
Stroke	78 (mm)
Compression ratio	11.9
Connection rod length	130 (mm)
Number of cylinders	4
Piston pin offset	0 (mm)
Displacement	1600 (cc)
Intake valve open	19 BTDC (deg)
Intake valve close	52 ABDC (deg)
Exhaust valve open	47 BBDC (deg)
Exhaust valve close	15 ATDC (deg)
Piston surface area	6600 (mm^2)
Cylinder surface area	7700 (mm^2)
Number of strokes	4

Table 3.
GDI engine specification.

$$\frac{dm_u u_u}{d\alpha} = -p_c \frac{dV_u}{d\alpha} - \sum \frac{dQ_{Wu}}{d\alpha} - h_u \frac{dm_B}{d\alpha} - h_{BB,u} \frac{dm_{BB,u}}{d\alpha} \tag{2}$$

where dm_u represents the change of the internal energy in the cylinder, $p_c \frac{dV}{da}$ is the piston work, $\frac{dQ_F}{da}$ stands for the fuel heat input, $\frac{dQ_W}{da}$ is wall heat loses, and $h_u \frac{dm_b}{da}$ represents the enthalpy flow from the unburned to the burned zone due to the conversion of a fresh charge to combustion products. The heat flux between the two zones is neglected. $h_{BB} \frac{dm_{BB}}{da}$ represents the enthalpy due to blow by, u and b in the subscript are unburned and burned gas.

Moreover, the sum of the zone volumes must be equal to the cylinder volume, and the sum of the volume changes must be equal to the cylinder volume change:

$$\frac{dV_b}{d\alpha} + \frac{dV_u}{d\alpha} = \frac{dV}{d\alpha} \tag{3}$$

$$V_b + V_u = V \tag{4}$$

The amount of burned mixture at each time setup is obtained from the Vibe function. For all other terms, for instance, wall heat losses, etc., models similar to the single zone models with an appropriate distribution on the two zones are used [30].

2.2.1 Mass fraction burned

The Wiebe function is widely used in internal combustion engine applications to describe the fraction of mass burned in the combustion chamber during the combustion process [31, 32].

To represent the mass fraction burned, the Wiebe function is chosen. The Wiebe function for mass fraction burned is shown by the equation below:

$$x_b = 1 - exp\left[-a\left(\frac{\theta - \theta_o}{\Delta\theta}\right)^{m+1}\right] \tag{5}$$

where x_b is mass fraction burned, θ_o is start of combustion, $\Delta\theta$ is burn duration, a is efficiency factor, and b is shape factor (a and b are constant factors depending on the fuel). For this research, a complete combustion is assumed. The value of "a" is chosen to be 6.9 and the value of "b" to be 3 [33, 34]. The shape factor affects ignition delay and initial flame development. However, the values are subject to further analysis and provide scope for future research.

2.2.2 Mass fraction burned

The heat transfer to the walls of the combustion chamber is calculated from

$$Q_{wi} = hA_i(T_c - T_{wi}) \tag{6}$$

where Q_{wi} is wall heat flow, A_i is surface area, h is heat transfer coefficient, T_c is temperature of gas in the cylinder, and T_{wi} is wall temperature.

The Woschni model is selected to determine the heat transfer coefficient [35].

2.3 A description of exhaust emission model

In AVL BOOST the model of formation on NOx is based on Pattas and Hafner [30], which incorporates the Zeldovich mechanism [36]. The reaction of Zeldovich mechanism is given in **Table 4**.

The rate of NOx production was obtained using Eq. (7):

$$r_{NO} = C_{PPM}C_{KM}(2,0).\left(1-\alpha^2\right)\left(\frac{r_1}{1+\alpha AK_2} + \frac{r_4}{1+AK_4}\right). \tag{7}$$

$$\alpha = \frac{C_{NO.act}}{C_{NO.equ}} \cdot \frac{1}{C_{KM}} \tag{8}$$

$$AK_2 = \frac{r_1}{r_2 + r_3} \tag{9}$$

$$AK_4 = \frac{r_4}{r_5 + r_6} \tag{10}$$

In the above equation, C_{PPM} represents post-processing multiplier, C_{KM} denotes kinetic multiplier, C stands for molar concentration in equilibrium, and r_i represents reactions rates of the Zeldovich mechanism.

	Stoichiometry	Rate $k_i = k_{0,i}T^a e^{\left(\frac{-TA_i}{T}\right)}$	K_0 (cm³/mol s)	a [−]	T_A [K]
R1	$N_2 + O = NO + N$	$r_1 = k_1c_{N2}c_o$	4.93E13	0.0472	38,048,01
R2	$O_2 + N = NO + O$	$r_2 = k_2c_{O2}c_N$	1.48E08	1.5	2859,01
R3	$N + OH = NO + H$	$r_3 = k_3c_{OH}c_N$	4.22E13	0.0	0,0
R4	$N_2O + O = NO + NO$	$r_4 = k_4c_{N2O}c_O$	4.58E13	0.0	12,130,6
R5	$O_2 + N_2 = N_2O + O$	$r_5 = k_5c_{O2}c_{N2}$	2.25E10	0.825	50,569,7
R6	$OH + N_2 = N_2O + H$	$r_6 = k_6c_{OH}c_{N2}$	9.14E07	1.148	36,190,66

Table 4.
Reactions according to the Zeldovich mechanism.

The model of NOx formation is based on Onorati et al. [37]:

$$r_{CO} = C_{Const}(r_1 + r_2).(1 - \alpha) \tag{11}$$

$$\alpha = \frac{C_{CO.act}}{C_{CO.equ}} \tag{12}$$

In Eq. (11), C represents molar concentration in equilibrium and r_i represents reaction rates based on the model.

The unburned HC have different sources. A complete description of HC formation still cannot be given, and the achievement of a reliable model within a thermodynamic approach is definitely prevented by the fundamental assumptions and the requirement of reduced computational times. Still, a phenomenological model which accounts for the main formation mechanisms and is able to capture the HC trends as function of the engine operating parameter may be proposed. The following important sources of unburned HC can be identified in SI engines [38]:

1. During the intake and compression stroke, fuel vapor is absorbed into the oil layer and deposits on the cylinder walls. The following desorption occurs when the cylinder pressure decreases during the expansion stroke, and complete combustion cannot take place anymore.

2. A fraction of the charge enters the crevice volumes and is not burned since the flame quenches at the entrance.

3. Occasional complete misfire or partial burning takes place when combustion quality is poor.

4. Quench layers on the combustion chamber wall which are left as the flame extinguishes prior to reaching the walls.

5. The flow of fuel vapor into the exhaust system during valve overlap in gasoline engines.

The first two mechanisms and in particular the crevice formation are considered to be the most important and need to be accounted for in a thermodynamic model. Partial burn and quench layer effect cannot be physically described in a quasi-dimensional approach, but may be included by adopting tunable semiempirical correlations.

The formation of unburned HC in the crevices is described by assuming that the pressure in the cylinder and in the crevices is the same and that the temperature of the mass in the crevice volumes is equal to the piston temperature.

The mass in the crevices at any time is described by Eq. (13):

$$m_{crevice} = \frac{pV_{crevice}M}{RT_{piston}} \tag{13}$$

In Eq. (13), $m_{crevice}$ represents the mass of unburned charge in the crevice, p denotes cylinder pressure, $V_{crevice}$ stands for total crevice volume, M represents unburned molecular weight, T_{piston} is the temperature of the piston, and R denotes gas constant.

The second important source of HC is the presence of lubricating oil in the fuel or on the walls of the combustion chamber. During the compression stroke, the fuel vapor pressure increases, so, by Henry's law, absorption occurs even if the oil was saturated during the intake. During combustion the concentration of fuel vapor in the burned gases goes to zero, so the absorbed fuel vapor will desorb from the liquid oil into the burned gases. Fuel solubility is a positive function of the molecular weight, so the oil layer contributed to HC emissions depending on the different solubilities of individual hydrocarbons in the lubricating oil.

The assumptions made in the development of the HC absorption/desorption are the following:

1. Fuel is constituted by a single hydrocarbon species, completely vaporized in the fresh mixture.

2. The oil film temperature is at the same as the cylinder wall.

3. Traverse flow across the oil film is negligible.

4. Oil is represented by squalane (C30H62), whose characteristics resemble those of the SAE5W20 lubricant.

5. Diffusion of the fuel in the oil film is the limiting factor, for the diffusion constant in the liquid phase is 104 times smaller than the corresponding value in the gas phase.

The radial distribution of the fuel mass fraction in the oil film can be determined by solving the diffusion Eq. (14):

$$\frac{\partial w_F}{\partial t} - D\frac{\partial^2 w_F}{\partial r^2} = 0 \tag{14}$$

In Eq. (14), w_F represents fuel's mass fraction in the oil film, t is the time, r stands for radial position in the oil film (distance from the wall), and D is relative (fuel-oil) diffusion coefficient.

3. Result and discussion for gasoline PFI engine

The present research focused on the performance and emission characteristics of the methanol-gasoline blends. Various concentrations of the blends 0% methanol (M0), 5% methanol (M5), 10% methanol (M10), 20% methanol (M20), 30% methanol (M30), 50% methanol (M50), and 85% methanol (M85) by volume were analyzed.

3.1 Engine performance characteristics

Figures 3 and **4** show the results of the brake power and torque for methanol-gasoline blended fuels at various engine speeds.

The variation of brake power versus engine speed was obtained at full load conditions for various concentrations of M5, M10, M20, M30, M50, and net gasoline M0. When the methanol content in the blended fuel was increased (M10, M20 and M30), the brake power slightly increased for all engine speeds. The brake power at net gasoline was higher in comparison of M50 for all engine speeds. The methanol's heat of evaporation is higher than that of gasoline fuel, providing air-fuel charge cooling and increasing the density of the charge. The methanol's blended fuel causes the equivalence ratio of blend approaches to stoichiometric condition which can lead to a better combustion. However, methanol's heating value is lower

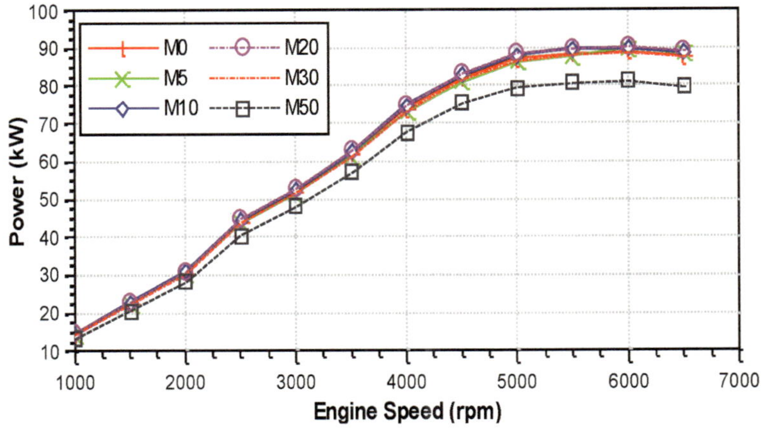

Figure 3.
Influence of methanol-gasoline blended fuels on brake power.

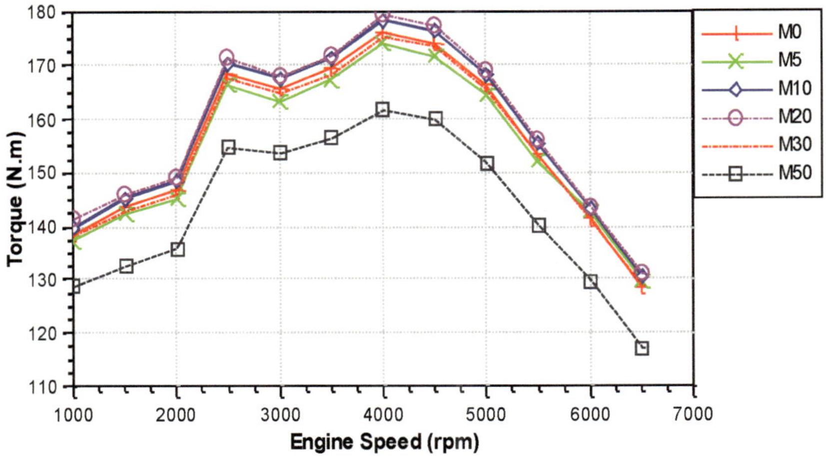

Figure 4.
Influence of methanol-gasoline blended fuels on engine torque.

than that of gasoline, and it can neutralize the previous positive effects. Consequently, a lower power output is obtained at M50.

Figure 4 shows the engine torque for various percentages of methanol in its blends with gasoline. Increasing the methanol content (M10 and M20) increased slightly the torque of the engine. The brake torque of gasoline was higher than those of M30 and M50.

Because of the existence of oxygen in the methanol chemical component, and the increase of methanol, lean mixtures are produced that decrease the equivalent air-fuel ratio to a lower value, and due to the presence of oxygen in the combustion chamber, the burning is more efficient.

Figure 5 shows the BSFC for various percentages of methanol in its blends with gasoline versus engine speeds. As shown in this figure, the BSFC increased as the methanol percentage increased. The reason has been known—the heating value and stoichiometric air-fuel ratio are the smallest for this fuel, which means that more fuel is needed for specific air-fuel equivalence ratio. The highest specific fuel consumption is obtained at M50 blended fuel.

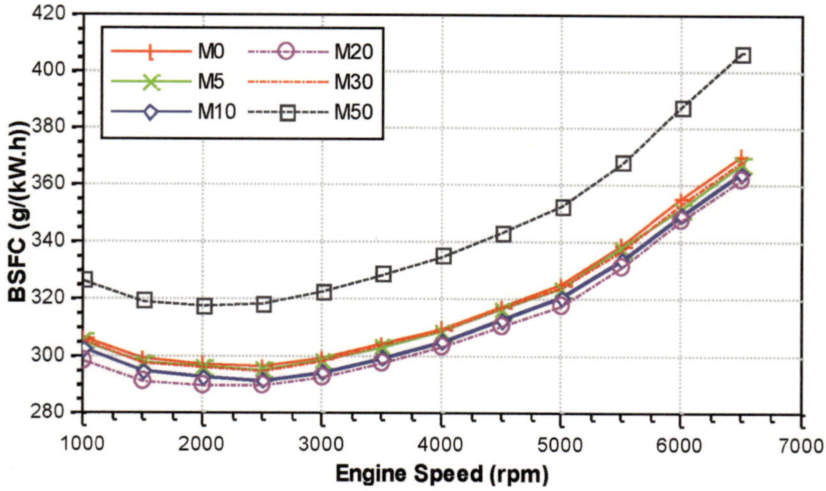

Figure 5.
Influence of methanol-gasoline blended fuels on brake-specific fuel consumption.

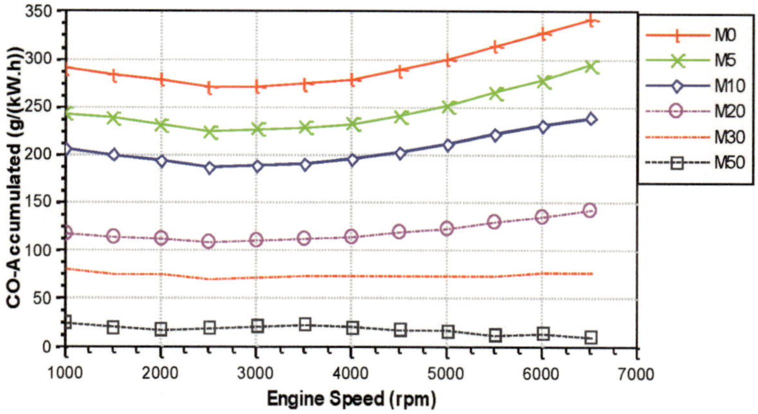

Figure 6.
Influence of methanol-gasoline blended fuels on CO emissions.

Furthermore, there is a small difference between the BSFC for net gasoline and the mixtures with methanol (M5 to M30). As engine speed increased reaching 2000 rpm, the BSFC decreased reaching its minimum value.

3.2 Emission characteristics

The effect of the methanol-gasoline blends on CO emissions versus engine speeds can be seen in **Figure 6**. When methanol percentage increases, the CO emissions decrease. The reason for this could be explained with the enrichment of oxygen owing to the methanol, in which an increase in the proportion of oxygen will promote the further oxidation of CO during the engine exhaust process. Another significant reason for this reduction is that methanol (CH_3OH) has less carbon than gasoline (C_8H_{18}).

The effect of the methanol-gasoline blends on HC emissions is shown in **Figure 7**. When methanol percentage increases, the HC concentration decreases. The HC emissions decrease with the increase of the relative air-fuel ratio.

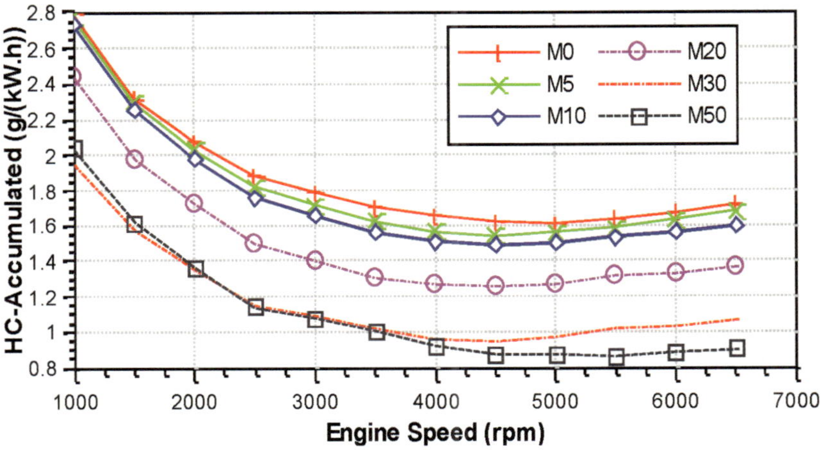

Figure 7.
Influence of methanol-gasoline blended fuels on HC emissions.

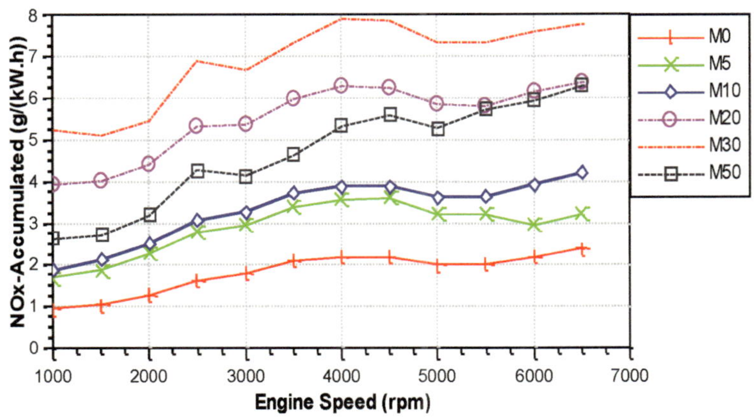

Figure 8.
Influence of methanol-gasoline blended fuels on NOx emissions.

The influence of the methanol-gasoline blends on NOx emissions can be seen in **Figure 8**. The methanol-gasoline blends lead to an increase in NOx emissions as compared to the net gasoline. When combustion process is closer to stoichiometric, flame temperature increases, and the NOx emissions increase as well.

4. Result and discussion for gasoline GDI engine

Different concentrations of the blends (0% methanol (M0), 5% methanol (M5), 10% methanol (M10), 20% methanol (M20), 30% methanol (M30), and 50% methanol (M50) by volume) were analyzed using AVL BOOST at full load conditions for GDI engine.

4.1 Engine performance characteristics

The results of the brake power and specific fuel consumption for methanol-gasoline blended fuels at different engine speeds are shown in **Figures 9** and **10**.

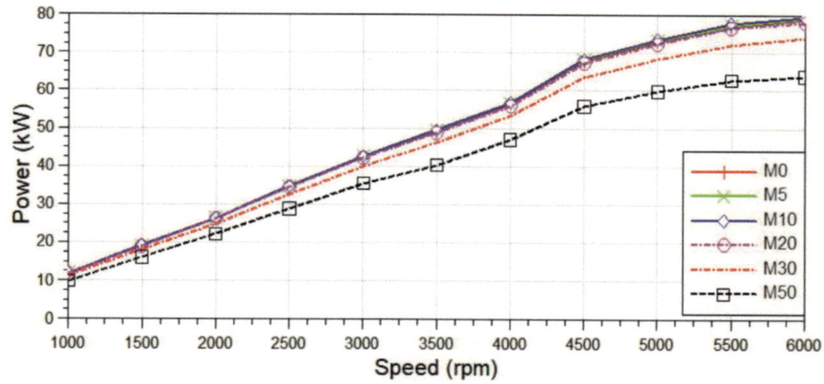

Figure 9.
Influence of methanol-gasoline blended fuels on brake power.

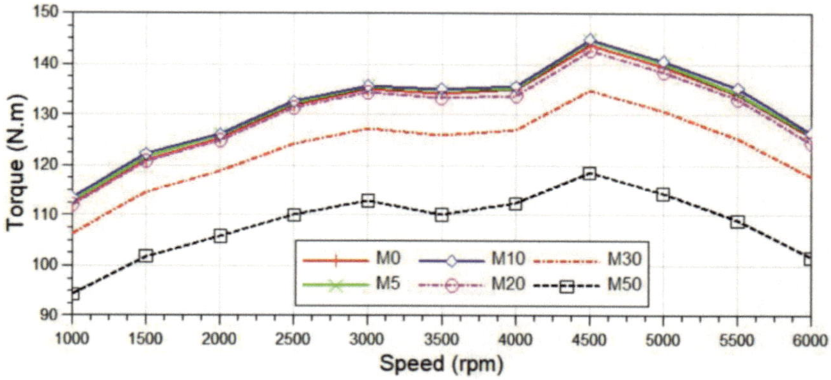

Figure 10.
Influence of methanol-gasoline blended fuels on engine torque.

The variation of brake power versus engine speed was obtained at full load conditions for various concentrations of M5, M10, M20, M30, M50, and net gasoline M0. When the methanol content in the blended fuel was increased (M5 and M10), there was not a significant increase in engine brake power. The engine brake power was higher in operation with gasoline in comparison to M50 for all engine speeds. The methanol's heat of evaporation is higher than that of gasoline fuel, providing air-fuel charge cooling and increasing the density of the charge. The methanol blended fuel causes the equivalence ratio of blend approaches to stoichiometric condition which can lead to a better combustion. However, methanol's heating value is lower than that of gasoline, and it can neutralize the previous positive effects. Consequently, a lower power output is obtained at M50.

Figure 10 shows the engine torque for various percentages of methanol in its blends with gasoline. Increasing methanol content (M5 and M10) increased slightly the torque of the engine. The brake torque of gasoline was higher than those of M20, M30, and M50.

Because of the existence of oxygen in the methanol chemical component, and the increase of methanol, lean mixtures are produced that decrease the equivalent air-fuel ratio to a lower value, and due to the presence of oxygen in the combustion chamber, the burning is more efficient.

Figure 11 shows BSFC for various percentages of methanol in its blends with gasoline versus engine speeds. As shown in this figure, the BSFC increased as the

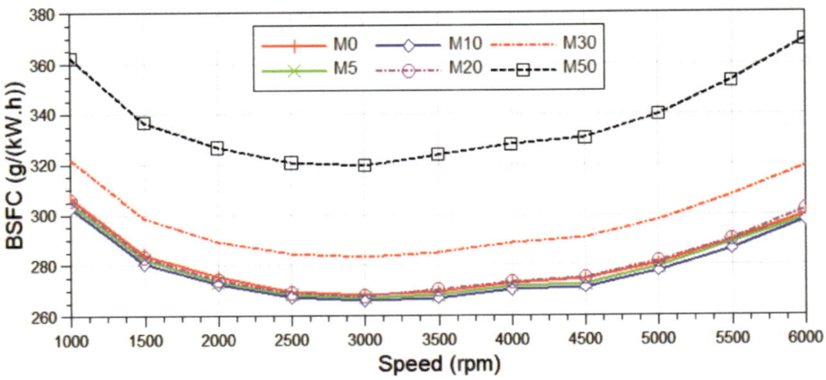

Figure 11.
Influence of methanol-gasoline blended fuels on brake-specific fuel consumption.

methanol percentage increased. This can be described with heating value and stoichiometric air-fuel ratio are the smallest for these two fuels, which means that for specific air-fuel equivalence ratio more fuel is needed. The highest specific fuel consumption is obtained at M50 blended fuel.

Furthermore, there is a small difference between the BSFC for net gasoline and the mixtures with methanol (M5 to M20). As engine speed increased reaching 3000 rpm, the BSFC decreased, reaching its minimum value.

4.2 Emission characteristics

The effect of the methanol-gasoline blends on CO emissions versus engine speeds can be seen in **Figure 12**. When methanol percentage increases, the CO emissions decrease. The reason for this could be explained with the enrichment of oxygen owing to the methanol, in which an increase in the proportion of oxygen will promote the further oxidation of CO during the engine exhaust process. Another significant reason for this reduction is that methanol (CH_3OH) has less carbon than gasoline (C_8H_{18}). Typical of direct injection engines is that they work with lean mixtures. Ensuring sufficient oxygen in the piston bowl for good flame

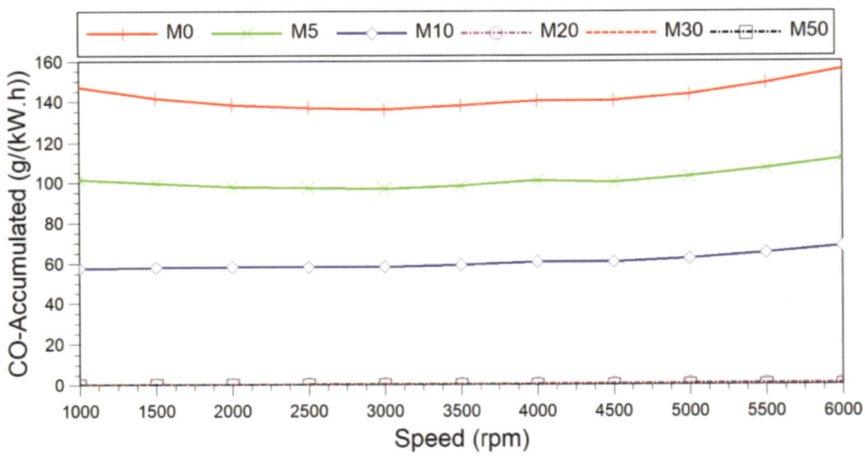

Figure 12.
Influence of methanol-gasoline blended fuels on CO emissions.

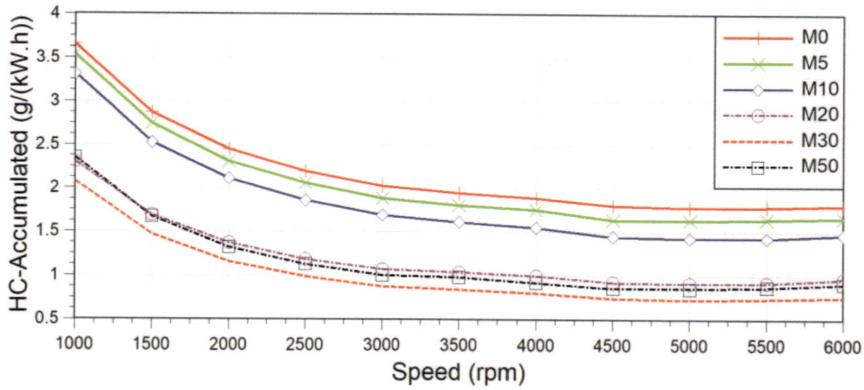

Figure 13.
Influence of methanol-gasoline blended fuels on HC emissions.

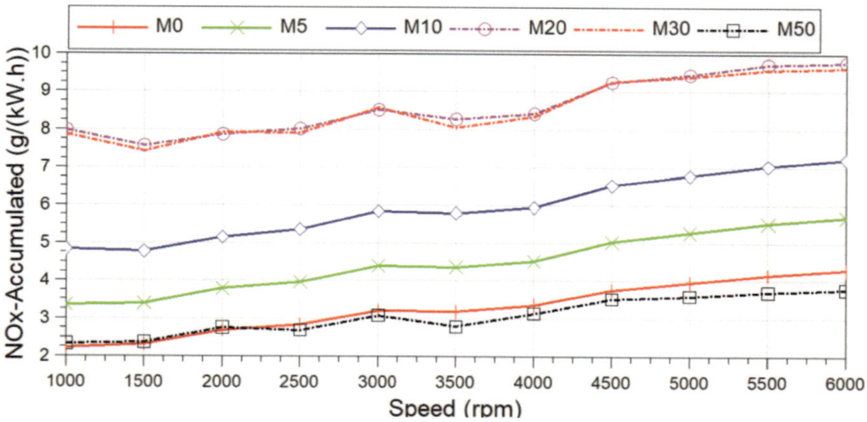

Figure 14.
Influence of methanol-gasoline blended fuels on NOx emissions.

propagation leads to higher combustion efficiency, reduced CO emissions, and optimal burn duration.

The influence of the methanol-gasoline blends on HC emissions is visible in **Figure 13**. When methanol percentage increases, the HC concentration decreases. The HC emissions decrease with the increase of the relative air-fuel ratio.

The effect of the methanol-gasoline blends on NOx emissions can be seen in **Figure 14**. When methanol percentage increases, the NOx concentration increases. Other authors [39] have obtained similar results. When combustion process is closer to stoichiometric, flame temperature increases, and the NOx emissions increase as well. The higher combustion temperature and local oxygen concentration in the peak temperature zone were the influencing factors for NOx emission formation.

5. Conclusions

The purpose of the present chapter is to demonstrate the influences of methanol addition to gasoline on a four-cylinder, four-stroke, multipoint injection system SI engine and a direct injection system SI on engine performance and emission characteristics. The summarized results from this study are the following.

With the increase of the percentage of methanol in the blended fuel, the engine brake power decreased for various engine speeds for PFI engine and GDI engine.

With the increase of the percentage of methanol in the blends (M5 to M10 for PFI engine and M5 to M10 for GDI engine), the brake power slightly increased, and with the increase of the methanol in the blends (M30 and M50), the brake power decreased.

As the percentage of methanol increased, the BSFC increased. The blended fuels show lower engine brake power and higher BSFC than net gasoline. Furthermore, there is a slight difference between the BSFC of gasoline and gasoline blended fuels (M10 and M20 for PFI engine and M5 and M10 for GDI engine).

When there is an increase in methanol percentage, the CO and HC concentration decreases for PFI and GDI engines.

Increasing the percentage of methanol in the blends leads to a significant increase in NOx emissions. The lowest NOx emissions are obtained at M50 methanol-gasoline blend at GDI engines.

Acknowledgements

The present chapter has been written with the Project No 2020-RU-03's financial assistance.

We are also eternally grateful to AVL-AST, Graz, Austria, for granting the use of AVL BOOST under the university partnership program.

Author details

Simeon Iliev
University of Ruse "Angel Kanchev", Ruse, Bulgaria

*Address all correspondence to: spi@uni-ruse

References

[1] Iliev S. Investigation of N-butanol blending with gasoline using a 1-D engine model. Proceedings of the 3rd International Conference on Vehicle Technology and Intelligent Transport Systems (VEHITS 2017). 2017:385-391. DOI: 10.5220/0006284703850391

[2] Pourkhesalian AM, Shamekhi AH, Salimi F. Alternative fuel and gasoline in an SI engine: A comparative study of performance and emissions characteristics. Fuel. 2010;**89**:1056-1063

[3] Varol Y, Oner C, Oztop HF, Altun S. Comparison of methanol, ethanol, or n-butanol blending with unleaded gasoline on exhaust emissions of an SI engine. Energy Source, Part A. 2014;**36**:938-948

[4] Chum H, Overend R. Biomass and renewable fuels. Fuel Processing Technology. 2001;**71**:187-195. DOI: 10.1016/S0378-3820(01)00146-1

[5] Egebäck K, Henke M, Rehnlund B, Wallin M, Westerholm R. Blending of ethanol in gasoline for spark ignition engines—Problem inventory and evaporative measurements. Rapport MTC. 2005;**5407**:8-9

[6] MAN Diesel &Turbo. Using Methanol Fuel in the MAN B&W ME-LGI Series, 5510-0172-00ppr; 2014

[7] Iliev S. Comparison of Ethanol and Methanol Blending with Gasoline Using Engine Simulation, Biofuels—Challenges and Opportunities. Mansour Al Qubeissi: IntechOpen; 2018. pp. 139-159. DOI: 10.5772/intechopen.81776

[8] Vancoillie J. Modeling the Combustion of Light Alcohols in Spark-Ignition Engines. Vakgroep Mechanica van Stroming, Warmte en Verbranding Sint-Pietersnieuwstraat 41, B-9000 Gent, Belgie: Ghent University; 2013. pp. 45-47

[9] Elfasakhany A. Investigation on performance and emissions characteristics of an internal combustion engine fueled with petroleum gasoline and a hybrid methanol–gasoline fuel. IJET-IJENS. 2013;**13**:24-43

[10] Akutsu Y, Toyoda F, Tomita K, Yoshizawa F, Tamura M, Yoshida T. Effect of exhaust from alcohol fuel on ozone formation in the atmosphere. Atmospheric Environment: Part A General Top. 1991:1383-1389. DOI: 10.1016/0960-1686(91)90247-5

[11] Liu D, Qi Q, Zhang H, et al. Properties, performance, and emissions of methanol-gasoline blends in a spark ignition engine. Proceedings of the Institution of Mechanical Engineers, Part D. 2005;**219**(3):405-412

[12] Pearson RJ, Turner JWG. GEM ternary blends: Testing iso-stoichiometric mixtures of gasoline, ethanol and methanol in a production flex-fuel vehicle fitted with a physical alcohol sensor. SAE Paper No. 2012-01-1279; 2012

[13] Silva R, Cataluna R, Menezes EW, Samios D, Piatnicki CMS. Effect of additives on the antiknock properties and Reid vapor pressure of gasoline. Fuel. 2005;**84**:951-959

[14] Raveendran K, Ganesh A. Heating value of biomass and biomass pyrolysis products. Fuel. 1996;**75**:1715-1720

[15] Shenghua L, Clemente ERC, Tiegang H, Yanjv W. Study of spark ignition engine fueled with methanol/gasoline fuel blends. Applied Thermal Engineering. 2007;**27**:1904-1910

[16] Bilgin A, Sezer I. Effects of methanol addition to gasoline on the performance and fuel cost of a spark ignition engine. Energy & Fuels. 2008;**22**:2782-2788

[17] Altun S, Oztop H, Oner C, Varol Y. Exhaust emissions of methanol and ethanol-unleaded gasoline blends in a spark ignition engine. Thermal Science. 2013;**17**(1):291-297

[18] Liu SH, Clemente ERC, Hu TG, et al. Study of spark ignition engine fueled with methanol/gasoline fuel blends. Applied Thermal Engineering. 2007;**27**(11–12):1904-1910. DOI: 10.1016/j.applthermaleng.2006.12.024

[19] Svensson E, Li C, Shamun S et al. Potential levels of soot, NOx, HC and CO for methanol combustion. SAE Technical Paper 2016-01-0887; 2016

[20] Liang B, Ge Y, Tan J, Han X, Gao L, Hao L, et al. Comparison of PM emissions from a gasoline direct injected (GDI) vehicle and a port fuel injected (PFI) vehicle measured by electrical low pressure impactor (ELPI) with two fuels: Gasoline and M15 methanol gasoline. Journal of Aerosol Science. 2013;**57**:22-31

[21] Iliev S. A comparison of ethanol and methanol blending with gasoline using a 1-D engine model. Procedia Engineering. 2015;**100**:1013-1022. DOI: 10.1016/j.proeng.2015.01.461

[22] Iliev S. A comparison of ethanol, methanol and butanol blending with gasoline and relationship with Eengine performances and emissions. In: Proceedings of the 29th DAAAM International Symposium; 2018. pp. 0505-0514

[23] Alla GA. Computer simulation of a four stroke spark ignition engine. Energy Conversion and Management. 2012;**43**(8):1043-1061

[24] Pourkhesalian AM, Shamekhi AH, Salimi F. Alternative fuel and gasoline in an si engine: A comparative study of performance and emissions characteristics. Fuel. 2010;**89**(5): 1056-1063

[25] Al-Baghdadi MAS. A simulation model for a single cylinder four-stroke spark ignition engine fueled with alternative fuels. Turkish Journal of Engineering and Environmental Sciences. 2007;**30**(6):331-350

[26] Gatowski J, Balles EN, Chun K, Nelson F, Ekchian J, Heywood JB. Heat Release Analysis of Engine Pressure Data, Technical Report. Warrendale, PA: Society of Automotive Engineers, Inc.; 1984

[27] Heywood JB. Engine combustion modeling—An overview. In: Mattavi JN, Amann CA, editors. Combustion Modeling in Reciprocating Engines. New York: Plenum Press; 1980. pp. 1-35. ISBN 0-306-40431-1

[28] Heywood JB. Internal Combustion Engine Fundamentals. New York: McGrawHill; 1988

[29] Iliev S, Hadjiev K. Theoretical study of engine performance working on methanol-gasoline blends. Proceedings of the WCE. 2014;**II**:1459-1463

[30] AVL List Gmbh. AVL Boost—Theory. Hans-List-Platz 1, A-8020 Graz, Austria; 2013

[31] Heywood JB et al. Development and use of a cycle simulation to predict SI engine efficiency and NOx emissions. SAE. 1979;**790291**

[32] Montenegro G, Onorati A, Piscaglia F, D'Errico G. Integrated 1D-multi-D fluid dynamic models for the simulation of ICE intake and exhaust systems. SAE Technical Paper Series 2007. pp. 3-7. Technical Report 2007-01-0495; 2007. DOI: 10.4271/2007-01-0495

[33] Ghojel J. Review of the development and applications of the Wiebe function: A tribute to the contribution of Ivan Wiebe to engine research. International

Journal of Engine Research. 2010;**11**(4): 297-312

[34] Agarwal AK, Karare H, Dhar A. Combustion, performance, emissions and particulate characterization of a methanol-gasoline blend (gasohol) fuelled medium duty spark ignition transportation engine. Fuel Processing Technology. 2014;**121**:16-24

[35] Woschni G. A Universally Applicable Equation for the Instantaneous Heat Transfer Coefficient in the Internal Combustion Engine, Technical Report; SAE Technical Paper; 1967

[36] Bowman C. Kinetics of pollutant formation and destruction in combustion. Progress in Energy and Combustion Science. 1975;**1**(1):33-45

[37] Onorati A, Ferrari G, D'Errico G. 1D unsteady flows with chemical reactions in the exhaust duct-system of S.I. engines: Predictions and Experiments. SAE Paper No. 2001-01-0939. 2001

[38] Montenegro G, Onorati A. Modeling of silencers for IC engine intake and exhaust systems by means of an integrated 1D-multiD approach. SAE International Journal of Engines. 2009; **1**(1):466. DOI: 10.4271/2008-01-0677

[39] Gong Li Y, Deng YW, Yuan WH, Fu J, Zhang B. Experimental comparative study on combustion, performance and emissions characteristics of methanol, ethanol and butanol in a spark ignition engine. Applied Thermal Engineering. 2017;**115**: 53-63. DOI: 10.1016/j. applthermaleng.2016.12.037

Chapter 5

Knock Suppression of a Spark-Ignition Aviation Piston Engine Fuelled with Kerosene

Enhua Wang, Chenyao Wang, Fujun Zhang, Huasheng Cui, Chuncun Yu, Bolan Liu, Zhenfeng Zhao and Changlu Zhao

Abstract

Spark-ignition (SI) engine has a high power density, making it suitable for unmanned aerial vehicles. Normally, gasoline fuel with a high octane number (ON) is used for a spark-ignition engine. However, gasoline fuel is easy to be evaporated and has a low flash point which is unsafe for aviation engines. Kerosene with a high flash point is safer than gasoline. In this chapter, the combustion characteristics of kerosene for a spark-ignition aviation piston engine are analyzed. A three-dimensional (3D) model is setup, and the combustion process of the engine fuelled with kerosene is simulated. Later, the knock limit extension by water injection is evaluated experimentally. The results indicate that water injection can suppress the knock of SI engine with kerosene in some extent and the output power can be improved significantly.

Keywords: aviation piston engine, spark ignition, kerosene, knock suppression, water injection

1. Introduction

Four-stroke spark-ignition (SI) piston engines have advantages of good fuel economy, high power to weight density, low noise, low cost, and easy maintenance, making them suitable for helicopters and unmanned aerial vehicles [1]. However, gasoline is volatile and easy to be ignited, which is not safe as aviation fuel. In contrast, kerosene is safer and has been used widely for airplanes. If gasoline can be replaced by kerosene on a four-stroke SI piston engine, the aviation safety will be improved significantly. Therefore, the combustion characteristics of four-stroke SI piston engine using kerosene as fuel need to be investigated.

The physical properties of kerosene are very different from gasoline. Kerosene has a lower volatility and higher viscosity. Accordingly, the spray penetration, spray velocity, and cone angle are different [2]. As a wide distillation fuel, the chemical properties of kerosene also differ from gasoline. Lots of experiments were conducted to investigate the combustion characteristics of kerosene [3, 4]. For example, the ignition delay time of kerosene was measured as a function of temperature, pressure, and equivalence ratio using a shock tube [5] or laser-induced fluorescence (LIF) imaging [6].

The combustion performance of kerosene has been investigated comprehensively for turbojet and scramjet engines [7, 8]. However, the operation conditions of aviation piston engine deviate from those of turbojets and scramjets significantly. Very few investigations focused on the combustion performance of kerosene for internal combustion engines [9–11]. Fernandes et al. studied the performance of a heavy-duty diesel engine fuelled with JP-8. The torque output and fuel consumption were similar with diesel fuel. However, the injection duration was enlarged to compensate the low fuel density, and the ignition delay time was increased because of a low cetane number. In addition, the emissions of nitrogen oxide (NOx) and particle material (PM) were decreased apparently [12]. Tay et al. developed a reaction mechanism composed of 48 species and 152 reactions for kerosene, and the accuracy was validated in an optical engine [13].

Generally, the octane number (ON) of gasoline is in the range of 70–97, whereas the octane number of kerosene is much lower as only 20–50. Therefore, the knock phenomenon of kerosene is much severe in an SI engine, which will keep the power and economy of the engine in a low level [14]. To control the knock of an SI engine, various methods can be adopted, for example, postponing ignition time, reducing compression ratio, and using antiknock additives. Recently, water injection got wide attentions to control the super knock of gasoline engine [15]. There are basically two methods: port water injection and in-cylinder direct water injection. When port water injection is employed, water is sprayed into the intake manifold. For direct water injection, water is injected directly into the combustion chamber. For port water injection, it is better to install the injector close to the intake valve. The knock suppression was increased for a mass ratio of water over fuel as 0.3 [16]. An experimental study based on a single-cylinder engine indicated that fuel with a lower octane number could be used if port water injection was installed [17]. Kim et al. performed an experiment, and water was sprayed into the cylinder in a pressure of 5 MPa using a gasoline direct injection (GDI) fuel injector. The knock suppression was observed evidently [18]. Wei et al. investigated the influences of water injection quantity on energy efficiency experimentally, and it was found that the energy efficiency maximized with a mass ratio of 0.15 [19].

For the knock suppression of SI aviation piston engine, Anderson et al. had studied the combustion performance of a Rotax 914 engine fuelled with kerosene blends. The octane numbers of the fuel blends were 87 and 70 via blending 100 ON aviation gasoline and JP-8. The brake mean effective pressure reduced evidently when the octane number of the fuel blends diminished. The effect of adjusting the ignition timing was very small for performance improvement [20]. Subsequently, the influences of mass fraction of JP-8 were investigated. The volume ratio of JP-8 changed from 85–27% blending with an 87 ON fuel. To keep a high engine power output, the volume ratio should be decreased accordingly [21]. Later, they used a pre-chamber jet ignition system to increase the flame propagation speed of kerosene, and the fuel octane number was decreased by about 10 [22].

It is critical to find methods that can suppress the knock for an aviation four-stroke SI engine with kerosene. However, very few investigations were performed currently. In this chapter, the knock suppression of an aviation four-stroke SI engine is investigated numerically at first. Then, port water injection is installed, and the experimental results for knock suppression are measured. The improvement of the indicated mean effective pressure (IMEP) is evaluated.

2. Mathematical method

The combustion of SI piston engine is mainly a turbulent flame propagation process. For the flow process in the cylinder, the mass, momentum, energy, and species conservation equations are modeled in the 3D numerical simulation model.

Mass:

$$\frac{\partial \rho}{\partial t} + \frac{\partial}{\partial x_i}(\rho u_i) = 0. \tag{1}$$

Momentum:

$$\frac{\partial}{\partial t}(\rho u_i) + \frac{\partial}{\partial x_i}(\rho u_i u_j) = -\frac{\partial p}{\partial x_i} + \mu \frac{\partial}{\partial x_i}\left(\frac{\partial u_i}{\partial x_j} + \frac{\partial u_j}{\partial x_i} - \frac{2}{3}\delta_{ij}\frac{\partial u_k}{\partial x_k}\right). \tag{2}$$

Energy:

$$\frac{\partial}{\partial t}(\rho h) + \frac{\partial}{\partial x_j}(\rho u_j h) = \frac{\partial p}{\partial t} + \frac{\partial}{\partial x_i}\left[\frac{\mu}{Pr}\frac{\partial h}{\partial x_j} + \frac{\mu}{Pr}\left(\frac{Pr}{Sc} - 1\right)\sum_{s=1}^{N} h_s \frac{\partial Y_s}{\partial x_j}\right]. \tag{3}$$

Species:

$$\frac{\partial}{\partial t}(\rho Y_s) + \frac{\partial}{\partial x_j}(\rho u_j Y_s) = \frac{\partial}{\partial x_j}\left(\frac{\mu}{Sc}\frac{\partial Y_s}{\partial x_j}\right) + \rho \dot{r}_s. \tag{4}$$

The density of the fluid is calculated by the following equation according to the equation of state for gaseous mixture.

$$\rho = p \Big/ \left[RT\sum_{s=1}^{N}(Y_s/M_s)\right]. \tag{5}$$

The enthalpy of the mixture can be calculated as

$$h = \sum_{s=1}^{N} Y_s h_s(T). \tag{6}$$

The above equations can be used in direct numerical simulation of turbulence, but it is difficult for the actual calculation process because the calculation load is extremely large. In this study, k - ε model is employed to simulate the turbulent flow in the cylinder. The turbulent eddy viscosity is defined as

$$\mu_t = C_\mu \bar{\rho} k^2/\varepsilon. \tag{7}$$

The transport equations for the turbulent kinetic energy k and the dissipation rate ε are expressed by

$$\bar{\rho}\tilde{u}_j \frac{\partial k}{\partial x_j} = \frac{\partial}{\partial x_j}\left[\left(\frac{\mu_t}{\sigma_k} + \mu\right)\frac{\partial k}{\partial x_j}\right] - \bar{\rho}\widetilde{u_i'' u_j''}\frac{\partial \tilde{u}_i}{\partial x_j} - \frac{\mu_t}{\bar{\rho}^2}\frac{\partial \bar{\rho}}{\partial x_i}\frac{\partial \bar{p}}{\partial x_i} - \bar{\rho}\varepsilon, \tag{8}$$

$$\bar{\rho}\tilde{u}_j \frac{\partial \varepsilon}{\partial x_j} = \frac{\partial}{\partial x_j}\left[\left(\frac{\mu_t}{\sigma_\varepsilon} + \mu\right)\frac{\partial \varepsilon}{\partial x_j}\right] - C_1\frac{\varepsilon}{k}\left[\bar{\rho}\widetilde{u_i'' u_j''}\frac{\partial \tilde{u}_i}{\partial x_j} + \frac{\mu_t}{\bar{\rho}^2}\frac{\partial \bar{\rho}}{\partial x_i}\frac{\partial \bar{p}}{\partial x_i}\right] - C_2\bar{\rho}\frac{\varepsilon^2}{k}. \tag{9}$$

The empirical constants for the standard k-ε model are assigned below: $C_1 = 1.44$, $C_2 = 1.92$, $C_\mu = 0.09$, $\sigma_k = 1.0$, and $\sigma_\varepsilon = 1.3$. A nonslip boundary condition is assumed regarding all solid surfaces of the computational domain, and the standard wall functions are used.

The chemical time scales are assumed being much smaller than the turbulent ones. Therefore, the coherent flame model is used for the turbulent premixed combustion of the kerosene. The mean turbulent reaction rate is defined as

$$\overline{\rho \dot{r}_{fu}} = -\rho_{fr} y_{fu,fr} S_L \Sigma, \tag{10}$$

where S_L is the mean laminar burning velocity and Σ is the flame surface density. The transport equation for the flame surface density is

$$\frac{\partial \Sigma}{\partial t} + \frac{\partial}{\partial x_j}\left(\overline{u}_j \Sigma\right) - \frac{\partial}{\partial x_j}\left(\frac{v_t}{\sigma_\Sigma}\frac{\Sigma}{\partial x_j}\right) = S_g - S_a + S_{LAM}, \tag{11}$$

where σ_Σ is the turbulent Schmidt number, v_t is the turbulent kinematic viscosity, S_g is the product of the turbulent strain rate and the flame surface, S_a is the annihilation of flame surface, and S_{LAM} is the contribution of laminar combustion.

The laminar flame speed can be computed as

$$S_L = S_{L0}\left(1 - 2.1 y_{EGR}\right)\left(\frac{T_{fr}}{T_{ref}}\right)^{a_1}\left(\frac{p}{p_{ref}}\right)^{a_2}, \tag{12}$$

where T_{ref} and p_{ref} are the reference values of the standard state and a_1 and a_2 are fuel-dependent parameters.

The water injected from the water injector forms small droplets and mixes with the gases in the cylinder. The spray model needs to consider the droplet movement, fragmentation, evaporation, and wall impingement. Considering the movement resistance and buoyancy of droplets in the gases, the following equation for the droplet velocity is obtained.

$$\frac{du_{id}}{dt} = \frac{3}{4} C_D \frac{\rho_g}{\rho_d} \frac{1}{D_d} \left|u_g - u_d\right| \left(u_{ig} - u_{id}\right) + \left(1 - \frac{\rho_g}{\rho_d}\right) g_i, \tag{13}$$

where u_{id} is the particle velocity vector, u_{ig} is the domain fluid velocity, C_D is the drag coefficient, D_d is the particle diameter, ρ_g and ρ_d are the densities of the gas and the droplet, and g_i is the gravitational acceleration vector.

The instantaneous droplet position vector can be determined by

$$\frac{dx_{id}}{dt} = u_{id}. \tag{14}$$

The evaporation process of the droplet is described by

$$m_d c_{p,d} \frac{dT_d}{dt} = \dot{Q}\left(1 + L\frac{\dot{f}_{vs}}{\dot{q}_s}\right). \tag{15}$$

Assuming the Lewis number is unity, the flux ratio can be written as

$$\frac{\dot{f}_{vs}}{\dot{q}_s} = \frac{-B_y}{h_\infty - h_s - \left(h_{vs} - h_{gs}\right)\left(Y_{v\infty} - Y_{vs}\right)}, \tag{16}$$

where

$$B_y = \frac{Y_{vs} - Y_{v\infty}}{1 - Y_{vs}}. \tag{17}$$

The convective heat flux supplied from the gas to the droplet is denoted by

$$\dot{Q} = D_d \pi \lambda Nu (T_\infty - T_s). \tag{18}$$

The KH-RT model is used for the breakup process of the droplets. The wavelength Λ and wave growth rate Ω of the fastest growing wave are

$$\Lambda = 9.02r \frac{\left(1 + 0.45Oh^{0.5}\right)\left(1 + 0.4T^{0.7}\right)}{\left(1 + 0.87We_g^{1.67}\right)^{0.6}}. \tag{19}$$

$$\Omega = \left(\frac{\rho_d r^3}{\sigma}\right)^{-0.5} \frac{0.34 + 0.38We_g^{1.5}}{(1 + Oh)\left(1 + 1.4T^{0.6}\right)}. \tag{20}$$

Then, the breakup time is calculated as

$$\tau_a = \frac{3.726C_2 r}{\Lambda\Omega}. \tag{21}$$

Rayleigh-Taylor disturbances are in continuous competition with Kelvin-Helmholtz surface waves. The RT mechanism is caused by quick deceleration of the droplets, resulting in a growth of surface waves. When the droplet diameter is larger than the wavelength of the critical disturbance wave, the process of droplet breakage caused by RT must be considered. The fastest growing frequency Ω_t and wave number K_t are

$$\Omega_t = \sqrt{\frac{2}{3\sqrt{3\sigma}} \frac{g_t|\rho_d - \rho_c|^{1.5}}{\rho_d + \rho_c}}, \tag{22}$$

$$K_t = \sqrt{\frac{g_t|\rho_d - \rho_c|}{3\sigma}}. \tag{23}$$

The corresponding wave length and breakup time are

$$\Lambda_t = C_4 \frac{\pi}{K_t}, \tag{24}$$

$$\tau_t = C_5 \frac{1}{\Omega_t}. \tag{25}$$

The ATKIM model is used for the ignition process in the cylinder, which considers the charge stratification, the available electrical energy, the heat losses to the spark plug, and the influences of turbulence on the early flame kernel.

The AnB model is used for the prediction of autoignition of the air-fuel mixture. The autoignition delay time of the combustible air-fuel mixture can be calculated as

$$\tau_d = A\left(\frac{ON}{100}\right)^{3.4017} p^{-n} e^{\frac{B}{T}}, \tag{26}$$

where ON is the octane number of kerosene and the parameters A, n, and B are calibrated by the measured data of the in-cylinder pressure.

3. Numerical simulation

The numerical model for the combustion process of Rotax 914 engine fuelled with kerosene is established in AVL Fire software based on the mathematical model. The combustion chamber is shown in **Figure 1(a)**. The original engine uses gasoline fuel, and a carburetor-type fuel supply system is installed. In order to ensure a reliable ignition, the combustion chamber employs a double spark plug arrangement. One inlet valve and one exhaust valve are used. Because kerosene is difficult to be evaporated, the fuel injection system is replaced by a port fuel injection (PFI) system in this study. The main technical parameters of the retrofitted engine are listed in **Table 1**. The 3D grid model of the combustion chamber is shown in **Figure 1(b)**. The function of automatic grid generation is adopted, and the grid size is set between 0.5 mm and 1.5 mm. The total number of grids is 363,785. After checking the quality of the grid, the numerical simulation is performed.

This chapter mainly studies the compression and combustion process in the cylinder. To simplify the calculation process, only the crankshaft angle ranging from −160° CA to 70° CA is simulated where both valves are closed. The piston moves between the top dead center (TDC) and the bottom dead center in the cylinder. The mesh of the volume swept by the piston is divided into five different mesh models according to the rotation angle of the crankshaft. Each mesh number is activated under a specified range of crankshaft angle. The mesh number of each

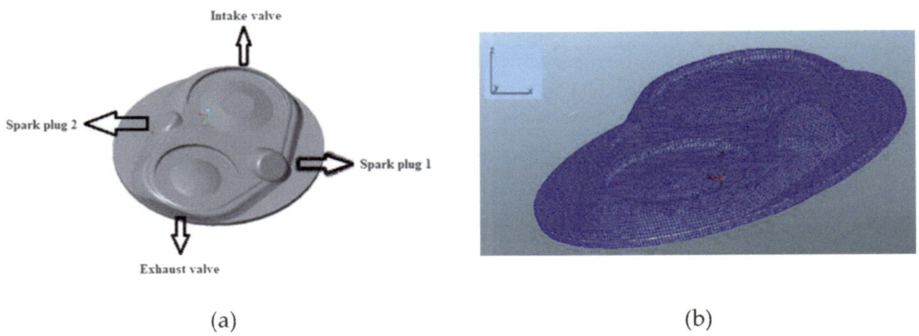

Figure 1.
Combustion chamber of the four-stroke spark-ignition engine: (a) 3D model; (b) volume mesh.

Item	Parameter	Unit
Displacement	1.211	L
Bore × stroke	79.5 × 61	mm
Cylinder number	4	—
Rated power	75	kW
Rated speed	5500	r/min
Max. torque	144	N.m
Speed at max. Torque	4900	r/min
Combustion chamber	Bathtub	
Valve actuation	2 valves per cylinder	
Air-fuel ratio	Stoichiometric	

Table 1.
Specifications of Rotax 914 engine.

mesh model remains unchanged, and only stretching or compression is experienced. The mesh quality in the process of tension or compression is checked. The initial pressure in the cylinder is set as 1.52 bar, and the initial temperature is set to 320 K. The temperatures of the cylinder head, cylinder wall, and piston top surface are set as 400, 380, and 450 K, respectively. The number of orifices for the water injector is six, and the diameter of the orifice is 2 mm.

In order to verify the accuracy of the numerical model, the in-cylinder pressure curve of the original engine with gasoline is simulated and compared with the test results, as shown in **Figure 2**. The engine speed is 5500 r/min and the throttle opening is 30%. The simulated in-cylinder pressure is in good agreement with the experimental one. Only the measured pressure near the TDC is slightly higher than the simulated one. The maximum relative error is 3.58%, which indicates that the numerical model has a high accuracy and can be used for the following simulation of the kerosene combustion process.

For the SI piston engine fuelled with kerosene, the effects of spray time, initial droplet size, and water quantity on knock suppression are analyzed using the established numerical model. The engine speed is set to 5500 r/min and the throttle opening is set to 30%. First, the effect of spray time on detonation suppression is studied. The injected kerosene quantity is fixed at 30 mg. The position of the injector is set in the center of the cylinder. The sprayed water temperature is 300 K and the initial droplet diameter is set to 0.1 mm. Only the starting time of water spray is changed and the other parameters are fixed. The start time of water injection is set as 120°, 80°, and 50° before TDC (BTDC) (represented by T1, T2, and T3, respectively).

The results for the in-cylinder pressure at three different injection times are shown in **Figure 3**. It can be seen that the closer the injection time is to the TDC, the higher the maximum burst pressure in the cylinder is. The useful work of the cycle ascends accordingly. However, the injection time has little effect on the in-cylinder pressure curve as a whole. The maximum pressure difference between T1 and T3 does not exceed 0.5 bar. T3 is close to the TDC and the in-cylinder temperature is higher than T1 and T2. The high temperature in the cylinder makes the liquid water easy to be evaporated. Part of the heat energy of the exhaust gas can be recovered. The combustion speed in the cylinder near the TDC is very fast. The injected water reduces the combustion rate of the combustible mixture. Since the crank angle of T3

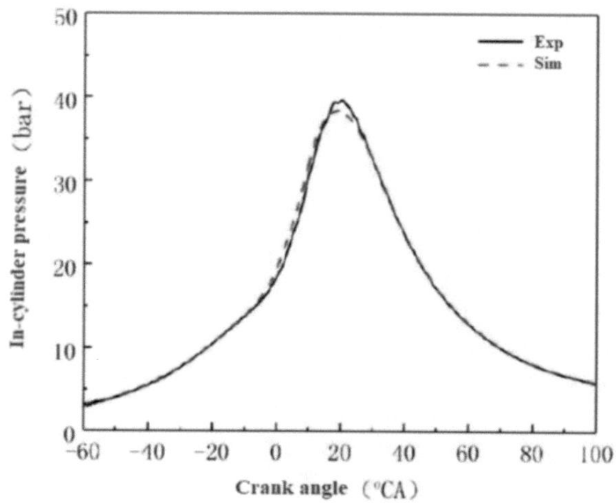

Figure 2.
Comparison of in-cylinder pressure between simulation and experimental data.

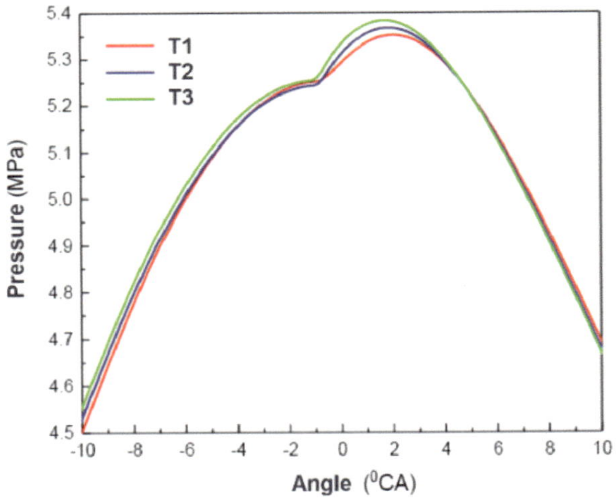

Figure 3.
Results of in-cylinder pressure with different water injection times.

is later than that of T1 and T2, the amount of water vapor in the cylinder is less, and the effect on the reaction rate reduction is the least significant. Therefore, the in-cylinder pressure of T3 is the highest. On the other hand, the closer the injection time is to the TDC, the earlier the phase of peak pressure appears. This is because the injected water vaporizes and absorbs heat in the cylinder and the oxygen concentration in the gas decreases, leading to a decrease of the reaction rate of the combustible mixture. The required compression work also decreases. The farther the injection time is away from the TDC, the longer the duration time of the combustion delay time in the cylinder. Therefore, the peak pressure with water injection time of T1 appears the maximum phase delay. The highest in-cylinder temperature for T1 is about 2300 K, while the highest temperature for T3 reaches 2500 K. The later the start time of water injection, the higher the temperature in the cylinder at the same crankshaft angle (**Figure 4**).

The influence of water injection time on the flame surface density is shown in **Figure 5**. The right spark plug starts to ignite at 52° CA BTDC, and the left spark plug ignites at 40° CA BTDC. The development of the double flames is observed. It can be seen that the flame surface density is greater at the same crank angle if the water injection time is closer to the TDC. The phase for the peak of the secondary heat release rate and the pressure rise rate is advanced accordingly. The peak phase of T3, T2, and T1 is 10°, 8°, and 4° CA BTDC, respectively.

In Fire, an index called combustion knock index (CKI) is used to label the knock intensity inside the cylinder. Generally, an obvious knock happens if the CKI is greater than 20. The results for the CKI under different injection times are shown in **Figure 6**. The CKI values for all the three injection times are maximized and exceed 20 at TDC. The CKI is the smallest for T1 and the strongest for T3. Meanwhile, the phase that the knock starts moves forward if the injection time is late. It can be seen that the knock happens at the position near the cylinder wall far away from the two spark plugs. The detonation phenomenon is reduced after the water injection. With the water injection time advanced from the TDC, the area where the detonation occurs tends to shrink, and the detonation intensity is weakened. This is due to the rapid vaporization of the water droplets after being sprayed into the cylinder, which absorbs a lot of heat in the cylinder. As a result, the in-cylinder temperature decreases. The effects of the compression and radiation of the burnt gas to the end gas reduce accordingly. Therefore, the autoignition time becomes longer than the

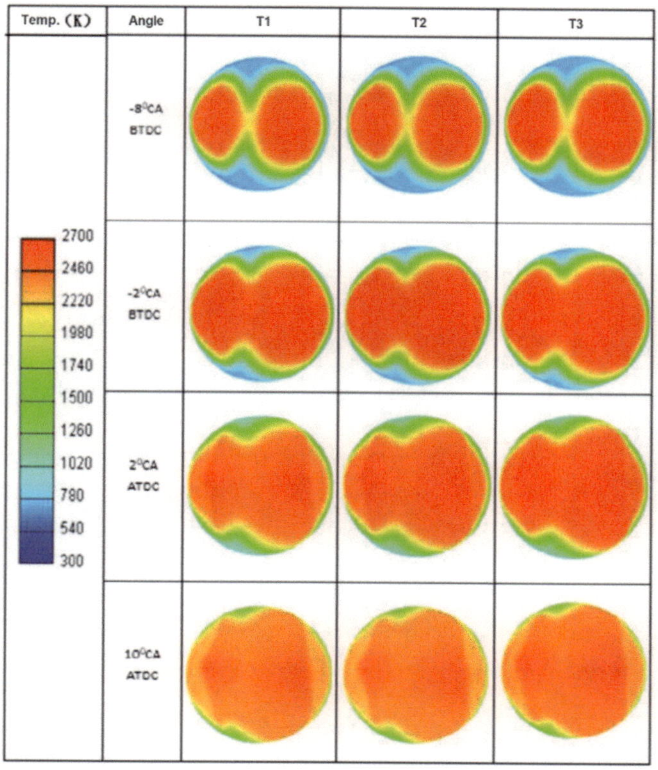

Figure 4.
Results of in-cylinder temperature with different water injection times.

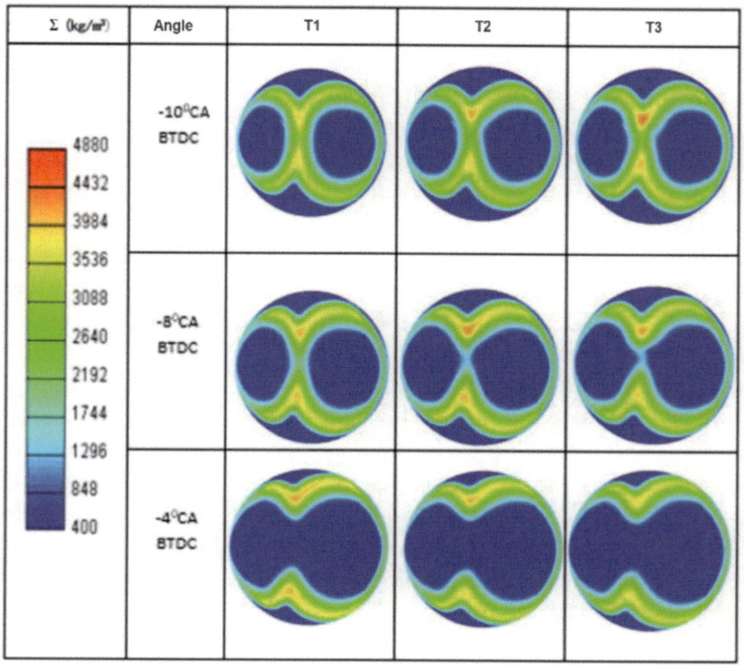

Figure 5.
Results of flame surface density with different water injection times.

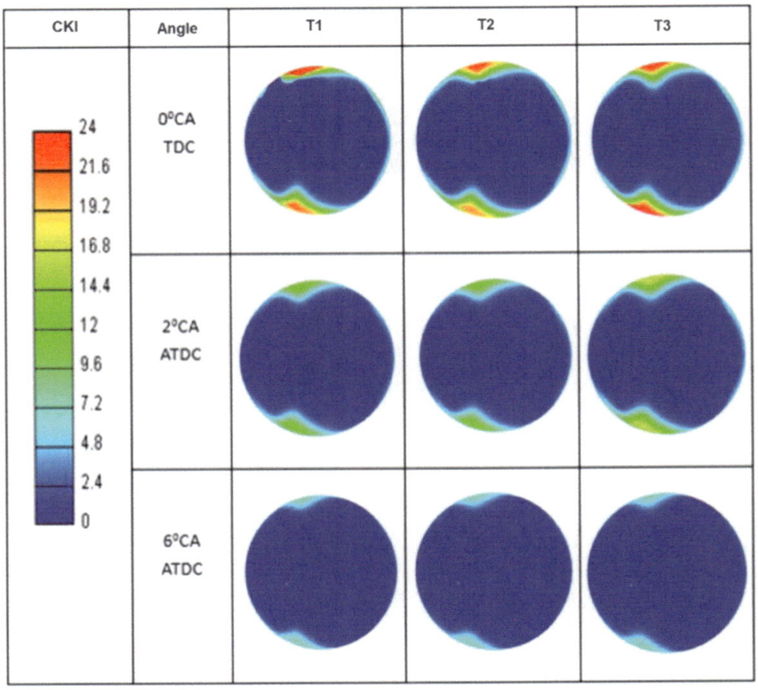

Figure 6.
Results for CKI with different water injection times.

flame propagation time, and the intensity of the detonation drops. Meanwhile, the thermal efficiency reduction caused by the heat loss of the knock phenomenon alleviates. For the aviation kerosene with an octane number of only 40–50, water injection can make the SI engine operate normally without power loss due to the reduced possibility of detonation. Meanwhile, the water injection time must be far enough away from TDC to suppress the knock effectively.

The initial droplet diameter of water spray will affect the subsequent droplet fragmentation and evaporation process. Therefore, it is necessary to study the effect of initial droplet diameter. Based on the above results, the water injection time is set to 120° CA BTDC. The water spray amount is set at 20 mg and the water temperature is set at 300 K. The initial diameters of the water droplets are set to 0.2, 0.1, 0.05, and 0.01 mm, respectively.

Figure 7 shows the influence of the initial droplet diameter on the in-cylinder pressure. The in-cylinder pressure becomes lower if the initial particle size is smaller. When the initial particle size is 0.01 mm, the in-cylinder pressure drops significantly. The in-cylinder peak pressure is about 5.5 MPa for an initial droplet diameter of 0.2 mm. However, when the initial particle size is 0.01 mm, the in-cylinder pressure has been reduced to 4.3 MPa, 21.8% lower than that with an initial droplet diameter of 0.2 mm. This is because the required energy acting on the water droplets to make them deformed and broken is weak for a small droplet diameter if the Weber number and the surface tension coefficient of the water droplet are fixed. When the initial diameter is too large, the energy is not enough to break the droplet, and only the deformation of the droplet occurs. Therefore, only a small part of the droplets with initial droplet diameters of 0.2 and 0.1 mm are broken. Accordingly, the amounts of water vapor are small. Therefore, their influence on the in-cylinder pressure is very limited.

For droplets with an initial diameter of 0.05 mm, there are many modes of fragmentation in the process of atomization. This effectively increases the contact

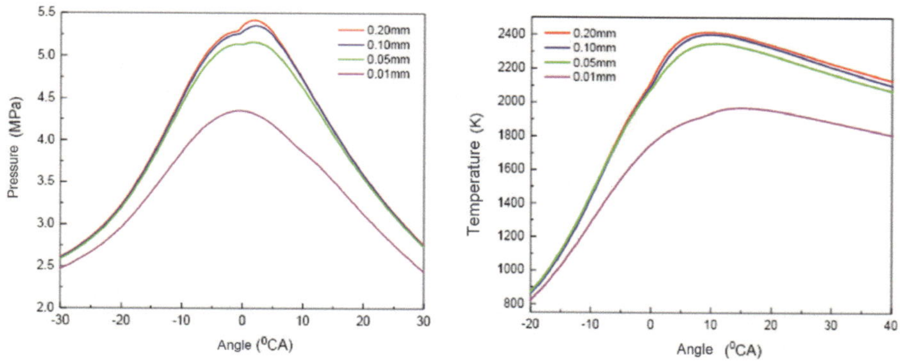

(a) (b)

Figure 7.
Results of the in-cylinder pressure (a) and temperature (b) with different initial water droplet diameters.

area between gas and liquid and accelerates the heat transfer rate. When the initial particle size decreases to 0.01 mm, the smaller droplet diameter makes the boundary conditions required for the fragmentation of the droplet smaller, and the critical Weber number becomes smaller. The required energy for the fragmentation and deformation of the droplets is lower. Therefore, the droplet is easier to be broken at this time. The increase of the surface area makes almost all the droplets evaporated, absorbing a lot of heat in the cylinder and reducing the heat release rate and the peak pressure of the combustion process.

The results for the in-cylinder temperature are shown in **Figure 7(b)**. When the initial droplet diameter is between 0.05 and 0.2 mm, the in-cylinder temperature changes slightly with a maximum temperature decrease up to 100 K. However, when the initial particle size is reduced to 0.01 mm, the in-cylinder temperature decreases greatly, and the maximum temperature reduction reaches nearly 500 K.

The results of the cloud chart for knock intensity with different initial droplet diameters are shown in **Figure 8**. The crank angle corresponding to the maximum knock intensity is different. The larger the initial droplet diameter is, the earlier the maximum detonation time will appear. The detonation intensity of the four particle sizes is also different. The larger the initial particle size is, the greater the CKI value of detonation is, indicating that the detonation is more dramatic. The knock phenomenon at the TDC with an initial particle size of 0.2 mm is the most severe, whose CKI value is as high as 30. When the initial droplet diameter is 0.1 mm, the detonation intensity maximizes at 1° CA ATDC, and the occurrence area is smaller. When the initial diameter is 0.05 mm, the maximum detonation happens at 2° CA ATDC, and the occurrence area is reduced apparently. The CKI value is decreased to around 20. When the initial droplet diameter is 0.01 mm, the detonation takes place at 10° CA ATDC. The corresponding CKI value drops to 12, which is reduced by 60% compared with an initial particle size of 0.2 mm.

The main purpose of spraying water into the cylinder is to reduce the temperature in the cylinder to suppress the knock. The amount of water injection directly affects the temperature reduction in the cylinder. Finally, the influence of water injection quantity is analyzed.

Figure 9(a) shows the results of the in-cylinder pressure under different water injection quantities. The in-cylinder pressure generally decreases with the increase of water spray quantity. The maximum burst pressure is 5.4 MPa when no water is sprayed. When the water injection quantity is 20 mg, the maximum burst pressure is 5.1 MPa, which is 5% lower than that without water injection. Furthermore, the

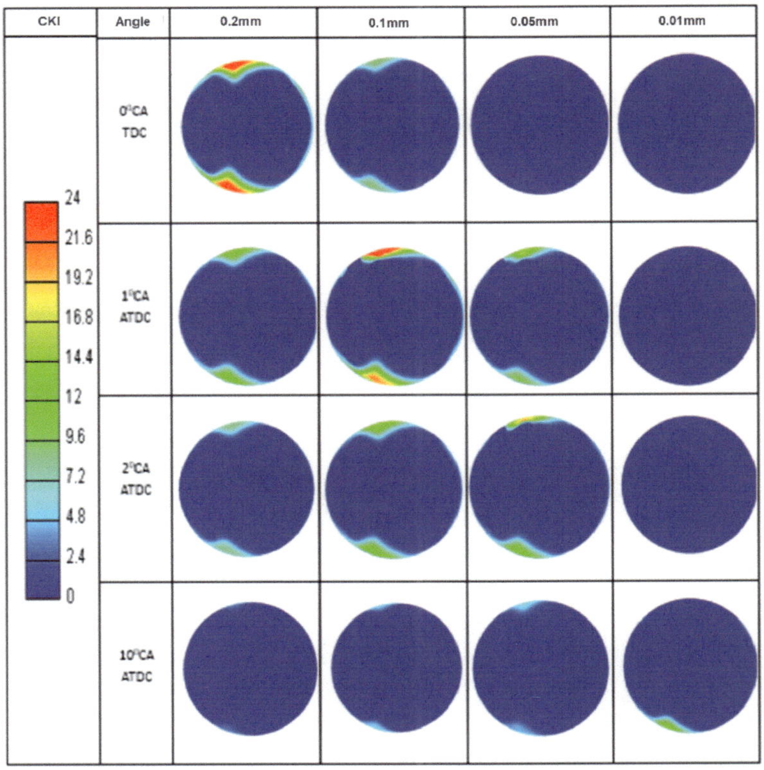

Figure 8.
Results of CKI with different initial water droplet diameters.

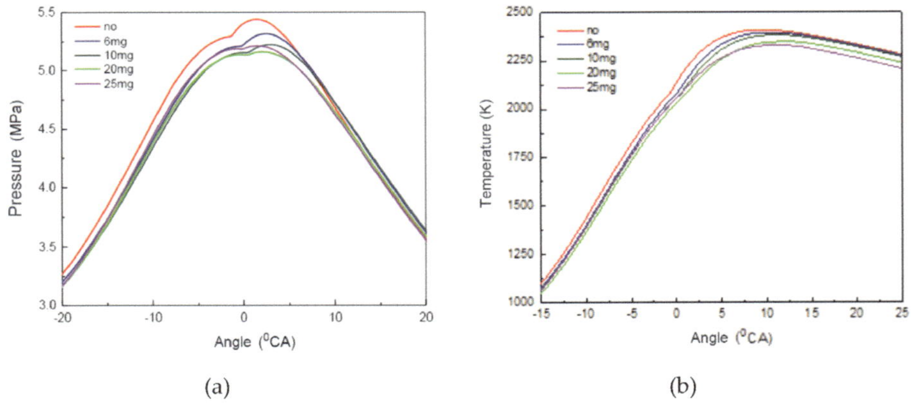

Figure 9.
Results of the in-cylinder pressure (a) and temperature (b) with different amounts of water injection.

larger the amount of water injection is, the later the phase of the maximum burst pressure appears.

Figure 9(b) gives a comparison of the in-cylinder temperature. When there is no water spray, the maximum average temperature in the cylinder is 2408 K. The temperature reduction in the cylinder is increased as the amount of water injection rises. When the water spray is 10 mg, the maximum temperature in the cylinder is 2365 K, which is 43 K lower than that without water spray. When the amount of the water spray is 25 mg, the maximum in-cylinder temperature drops to 2297 K, which is 111 K lower than that without water spray. In the case of water injection, the rise

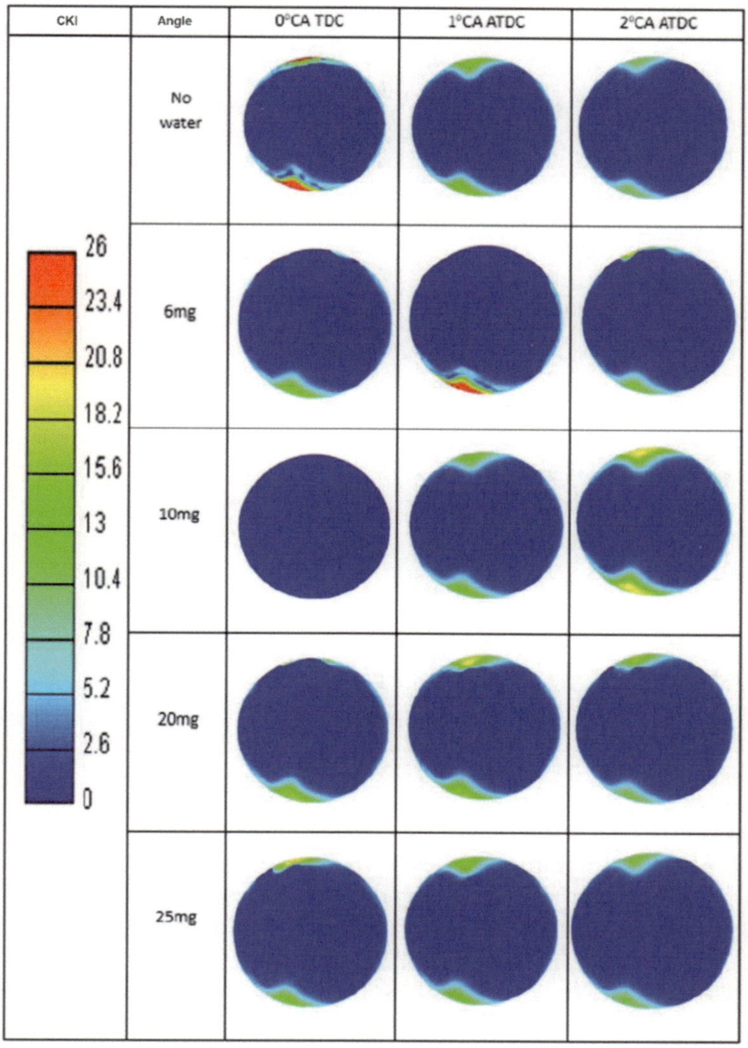

Figure 10.
Results of CKI with different initial water droplet diameters.

rate of the in-cylinder temperature slows down obviously. This is because after injection, water absorbs the heat in the cylinder and reduces the temperature.

The results for the cloud charts of CKI value with different water amounts are shown in **Figure 10**. The maximum CKI value decreases with the increase of water spray. The CKI value without water injection is as high as 26. When the water injection amount is 20 mg, the CKI value is reduced to 18, which is 31% lower than that without water injection.

4. Experimental results

To explore the effect of water injection on knock suppression of SI engine fuelled with kerosene, an engine experiment is carried out [23]. First, the original Rotax 914 engine was modified, and the engine test rig was built, as shown in **Figure 11**. The fuel supply system is first transformed into a PFI kerosene supply

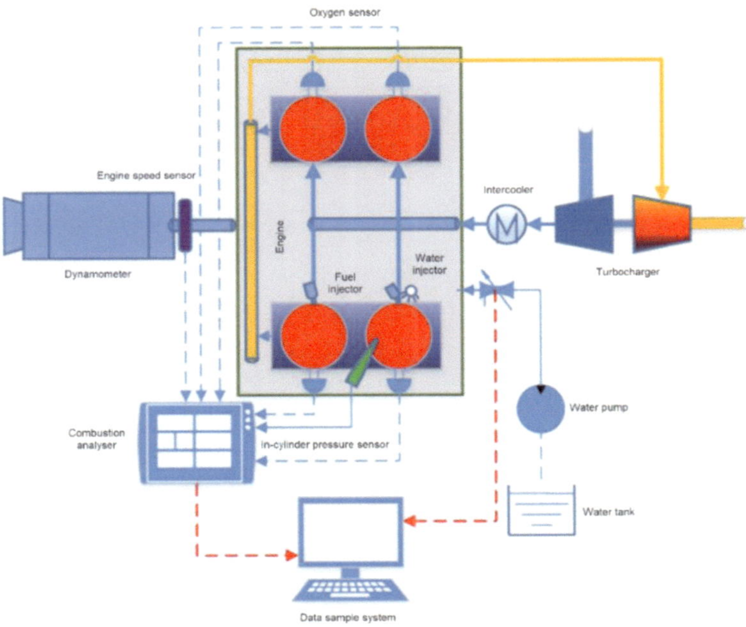

Figure 11.
Schematic of the engine test bench for the SI engine fuelled with kerosene together with water injection.

system. Meanwhile, a water injection system is added, which includes a water injector, a water pump, a water tank, and related pipelines. The water injector is installed at the intake manifold. Since the flame propagation speed of kerosene is significantly lower than that of gasoline, the ignition system of the original engine is modified. One of the spark plugs is replaced with a jet ignition device, as shown in **Figure 12**. The jet combustion pre-chamber uses the small orifices on the surface to inject the jet flame into the main combustion chamber and forms strong turbulent flames. As a result, the flame propagation distance is shortened, and the combustion speed is effectively accelerated. The volume of the jet ignition pre-chamber is 1.5 ml, and the diameter of the six holes on the surface is 1.5 mm.

The output power of the engine is absorbed by an eddy current dynamometer. During the test, kerosene is supplied to the first and third cylinders of the engine, and the original gasoline supply system is used for the second and fourth cylinders. A pressure sensor is installed in the first cylinder to measure the in-cylinder

Figure 12.
Photo of the jet ignition pre-chamber.

pressure. An engine speed sensor is installed at the crankshaft end. The collected data are sent to the combustion analyzer to calculate the heat release rate of the combustion process. **Table 2** lists the main equipment and relative accuracy of the test system.

When the engine speed is 5500 r/min and the throttle opening is 30%, the duration of water injection is set to 0, 2, and 4 ms, respectively. The results of the in-cylinder pressure are shown in **Figure 13(a)**. In the case of no water spray, the maximum pressure in the cylinder reaches 33.9 bar and fluctuates apparently near the TDC with an amplitude of 7.53 bar, indicating that the unburnt end gas in the cylinder is self-ignited and the knock phenomenon occurs. With water injection of 2 ms, the knock phenomenon has been alleviated, and the maximum in-cylinder pressure is increased from 33.9 to 35.4 bar. When the duration of water injection is increased from 2 to 4 ms, the pressure fluctuation in the cylinder caused by kerosene detonation disappears completely, while the maximum burst pressure in the cylinder decreases to 32.5 bar. The corresponding IMEP values are shown in **Figure 13(b)**, which also indicates that the in-cylinder pressure decreases significantly with the increase of water injection quantity.

Subsequently, the limit of knock suppression by water injection is studied under three different engine speeds of 4500, 5000, and 5500 r/min. For each engine speed, the throttle opening and the kerosene injection quantity as well as water injection quantity are increased continuously. The data are recorded until the maximum water injection duration is arrived. **Figure 14(a)** shows the comparison of kerosene injection quantity. In the case of no water injection, the maximum kerosene injection quantity under the three different engine speeds is about 20 mg. If the injection quantity exceeds this value, there will be an obvious detonation. The

Equipment	Model	Accuracy
Eddy current dynamometer	China Kaimai CW 160	<±0.5%
Fuel meter	Japan Ono MF-2200	<1%
Cylinder pressure sensor	Kistler 6115	<±0.5%
Lambda sensor	Bosch LSU4.9	<1%
Combustion analyzer	Kibox 2893	<1%

Table 2.
Main facilities of the experimental system.

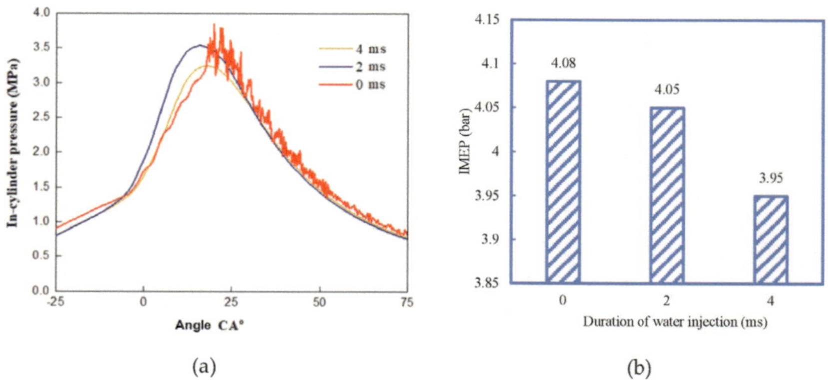

Figure 13.
Results for the in-cylinder pressure (a) and IMEP (b) with different water injection durations.

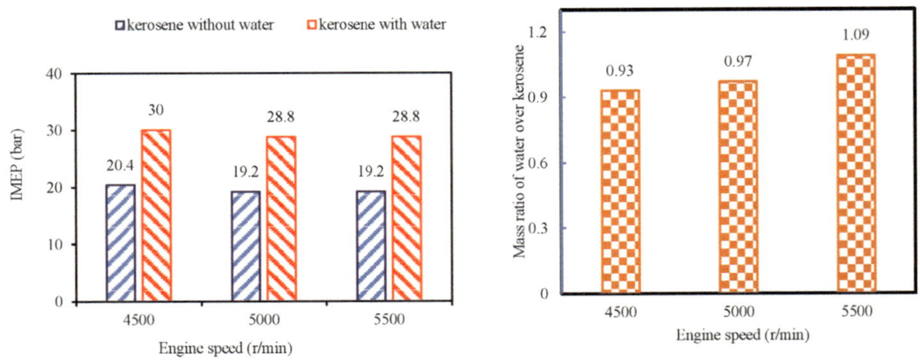

(a) (b)

Figure 14.
Results for the maximum amount of injected kerosene with/without water injection under different engine speeds (a) and the mass ratio of water over kerosene (b).

maximum kerosene injection quantity is increased by about 50% with water injection. The corresponding mass ratios of the sprayed water over kerosene are shown in **Figure 14(b)**. It can be seen that the mass ratio is close to 1:1 at three engine speeds.

Figure 15 shows the comparison of IMEP improvement under three different engine speeds. The IMEP can be increased from less than 8 to nearly 10 with water injection for each engine speed. The results show that using water injection can improve the kerosene injection quantity, leading to an increase of the IMEP and the power output of the engine. This is because after spraying water, the liquid water evaporates and absorbs heat, and the temperature in the cylinder decreases. Accordingly, the compression and radiation effect of the burnt gas imposed on the unburnt end gas weakens, and the autoignition time becomes longer, avoiding the violent detonation. The allowable charge of fresh mixture in the cylinder increases. Thus, more fuel can be injected and the engine's output power is improved. Meanwhile, it can be seen that the IMEP of the original engine with gasoline can reach 13 bar, and the IMEP of the kerosene fuel with water spray still needs to be enhanced.

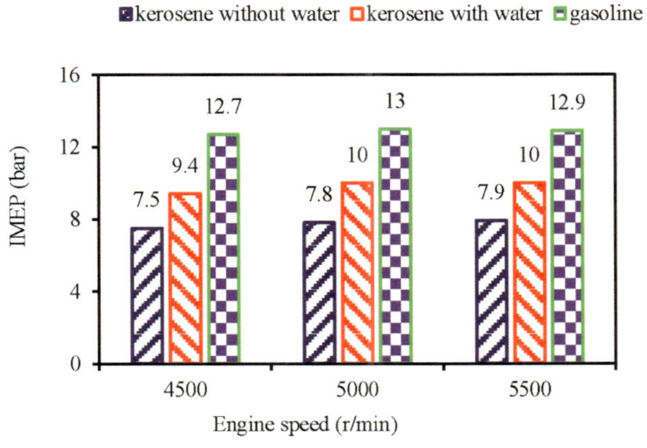

Figure 15.
Maximum improvements of the IMEP for the engine fuelled with kerosene together with water injection.

5. Conclusions

In this chapter, the feasibility of knock suppression using water injection for an SI engine fuelled with kerosene is investigated. First, a 3D numerical model is established in AVL Fire software. The influences of water injection timing, initial droplet diameter, and water injection quantity are analyzed. Then, an engine experiment is performed to evaluate the limit of IMEP improvement.

The results indicate that water injection can suppress the knock of kerosene effectively. The water injection timing should be advanced and far away from TDC because the evaporation of water droplets needs time. Accordingly, a lower pressure rise rate can be obtained, and the phase angle of the peak pressure can be postponed. The autoignition time enlarges, the flame propagation speed drops, and the intensity of the detonation decreases, due to a reduction of in-cylinder temperature. Meanwhile, a better result of knock suppression can be obtained with a smaller initial water droplet diameter, owing to a larger evaporation surface area of droplets. With the increase of the water injection amount, the reduction of in-cylinder temperature increases. The allowable throttle opening and injected kerosene amount also increase, resulting in a significant increase of the IMEP. The experimental results show that the maximum IMEP is improved from about 8 to 10 bar with an engine speed of 5500 r/min. However, an excessive water injection will lead to a decrease of the IMEP. To evaluate the effect of water injection comprehensively, an optimization design of the water injection system and the combustion chamber geometry is required. Meanwhile, more experimental work should be carried out, especially for the practical operation conditions of an unmanned aerial vehicle.

Acknowledgements

The authors would like to thank for the support of the National Natural Science Foundation of China (Grant No. 51876009).

Conflict of interest

The authors declare no conflict of interest.

Acronyms

3D	three-dimensional
SI	spark ignition
ATDC	after top dead center
BTDC	before top dead center
CKI	combustion knock index
GDI	gasoline direct injection
IMEP	indicative mean effective pressure
LIF	laser-induced fluorescence
ON	octane number
PFI	port fuel injection
PM	particle material
NOx	nitrogen oxide
TDC	top dead center

Author details

Enhua Wang*, Chenyao Wang, Fujun Zhang, Huasheng Cui, Chuncun Yu,
Bolan Liu, Zhenfeng Zhao and Changlu Zhao
School of Mechanical Engineering, Beijing Institute of Technology, Beijing, China

*Address all correspondence to: enhua.wang@yahoo.com

References

[1] Rozenkranc M, Ernst J. Tactical UAV engines integration in IAI. In: 2nd AIAA "Unmanned Unlimited" Systems, Technologies, and Operations — Aerospace; 15-18 September 2003, San Diego, California

[2] Yu W, Yang W, Zhao F. Investigation of internal nozzle flow, spray and combustion characteristics fueled with diesel, gasoline and wide distillation fuel (WDF) based on a piezoelectric injector and a direct injection compression ignition engine. Applied Thermal Engineering. 2017;**114**: 905-920

[3] Zeng W, Li H, Chen B, Ma H. Experimental and kinetic modeling study of ignition characteristics of Chinese RP-3 kerosene. Combustion Science and Technology. 2015;**187**: 396-409

[4] Cooke JA, Bellucci M, Smooke MD, Gomez A, Violi A, Faravelli T, et al. Computational and experimental study of JP-8, a surrogate, and its components in counterflow diffusion flames. Proceedings of the Combustion Institute. 2005;**30**:439-446

[5] Shao J, Zhu Y, Wang S, Davidson DF, Hanson RK. A shock tube study of jet fuel pyrolysis and ignition at elevated pressures and temperatures. Fuel. 2018; **226**:338-344

[6] Burkert A, Paa W. Ignition delay times of single kerosene droplets based on formaldehyde LIF detection. Fuel. 2016;**167**:271-279

[7] Felden A, Esclapez L, Riber E, Cuenot B, Wang H. Including real fuel chemistry in LES of turbulent spray combustion. Combustion and Flame. 2018;**193**:397-416

[8] Fan X, Yu G, Li J, Zhang X. Investigation of vaporized kerosene injection and combustion in a supersonic model combustor. Journal of Propulsion and Power. 2006;**22**:103-110

[9] Teoh YH, Masjuki HH, Kalam MA, Amalina MA, How HG. Impact of premixed kerosene fuel on performance, emission and combustion characteristics in partial HCCI engine. Energy Procedia. 2014;**61**:1830-1834

[10] Tay KL, Yang W, Zhao F, Yu W, Mohan B. A numerical study on the effects of boot injection rate-shapes on the combustion and emissions of a kerosene-diesel fueled direct injection compression ignition engine. Fuel. 2017; **203**:430-444

[11] Tay KL, Yang W, Li J, Zhou D, Yu W, Zhao F, et al. Numerical investigation on the combustion and emissions of a kerosene-diesel fueled compression ignition engine assisted by ammonia fumigation. Applied Energy. 2017;**204**:1476-1488

[12] Fernandes G, Fuschetto J, Filipi Z, Assanis D, McKee H. Impact of military JP-8 fuel on heavy-duty diesel engine performance and emissions. Proceedings of the Institution of Mechanical Engineers, Part D. 2007; **221**(8):957-970

[13] Tay KL, Yang W, Mohan B, An H, Zhou D, Yu W. Development of a robust and compact kerosene–diesel reaction mechanism for diesel engines. Energy Conversion and Management. 2016;**108**: 446-458

[14] Wang Z, Liu H, Reitz RD. Knocking combustion in spark-ignition engines. Progress in Energy and Combustion Science. 2017;**61**:78-112

[15] Hoppe F, Thewes M, Baumgarten H, Dohmen J. Water injection for gasoline engines: Potentials, challenges, and

solutions. International Journal of Engine Research. 2016;**17**(1):86-96

[16] Battistoni M, Grimaldi CN, Cruccolini V, Discepoli G. Assessment of port water injection strategies to control knock in a GDI engine through multi-cycle CFD simulations. SAE Technical Paper 2017-24-0034

[17] Miganakallu N, Naber JD, Rao S, Atkinson W, Barros S. Experimental investigation of water injection technique in gasoline direct injection engine. In: Proceedings of the ASME 2017 Internal Combustion Engine Division Fall Technical Conference; October 15-18 2017. Seattle, Washington, USA

[18] Kim J, Park H, Bae C, Choi M, Kwak Y. Effects of water direct injection on the torque enhancement and fuel consumption reduction of a gasoline engine under high-load conditions. International Journal of Engine Research. 2016;**17**(7):795-808

[19] Wei M, Sa NT, Turkson RF, Liu J, Guo G. Water injection for higher engine performance and lower emissions. Journal of the Energy Institute. 2017;**90**(2):285-299

[20] Anderson EK, Brown AC, Baranski J, Hoke JL. Performance of low-octane fuels in a Rotax 914 engine with advanced knock mitigation strategies. In: 47th AIAA/ASME/SAE/ASEE Joint Propulsion Conference & Exhibit; 31 July–3 August 2011. San Diego, California

[21] Baranski J, Anderson E, Grinstead K, Hoke J, Litke P. Control of fuel octane for knock mitigation on a dual-fuel spark-ignition engine. SAE Technical Paper 2013-01-0320

[22] Anderson EK, Attard WP, Brown A, Litke P, Grinstead K, Hoke J. Experimental study of a pre-chamber jet igniter in a turbocharged Rotax 914 aircraft engine. SAE Technical Paper 2013-01-1629

[23] Wang C, Zhang F, Wang E, Yu C, Gao H, Liu B, et al. Experimental study on knock suppression of spark-ignition engine fuelled with kerosene via water injection. Applied Energy. 2019;**242**: 248-259

Section 3

Improvements in Heavy-Duty Vehicle Systems

Chapter 6

Application of Variable Compression Ratio VCR Technology in Heavy-Duty Diesel Engine

Paweł Woś, Krzysztof Balawender, Mirosław Jakubowski, Artur Jaworski, Paulina Szymczuk and Adam Ustrzycki

Abstract

The chapter presents the application of variable compression ratio VCR technology for an internal combustion engine. The engine's overall construction has been designed as the deep reconfiguration of liquid-cooled, heavy-duty diesel engine of type IFA 4 VD 14,5/12-1 SRW. An eccentric shaft-based sliding mechanism has been used for changing position of engine cylinder head assembly. By ensuring a high stiffness of the engine layout together with additional sliding mechanism, the combustion chamber volume of the engine can be changed precisely without changing the displacement of the cylinders. In result, the engine compression ratio can be continuously varied during normal engine operation from the value of 19:1 down to 9:1.

Keywords: heavy-duty diesel engine, compression ratio, variable compression ratio technology, eccentric mechanism, internal combustion engine

1. Introduction

The desire to limit the adverse external effects associated with the development of motorization and transport, mainly based on the reciprocating internal combustion engine as the primary power source, reveals primarily with limitation the allowable level of emissions of harmful substances in the exhaust gases, that is, carbon monoxide, CO; hydrocarbons, HC; nitrogen oxides, NOx; and particulate matter, PM. Moreover, in the perspective of growing global warming phenomenon and dropping resources of liquid fuels, particular attention is paid to the reduction of fuel consumption and thus CO_2 emission. Unfortunately, the physical and chemical rules of working processes in piston engines do not allow to meet all the above requirements in a simple manner. Despite significant progress made in recent years, further improvement of the ecological and energetic parameters of reciprocating engines requires further changes in their design, covering practically all functional systems.

One of the more effective paths, although at the current stage of development is still having many technical and operational difficulties, is the construction of engines in which the compression ratio becomes a regulation parameter and can be

 IntechOpen

changed continuously in a wide range and relatively short time. In a conventional engine, the compression ratio—the ratio of the smallest to the largest cylinder volume at piston top dead center positions—is constant and determined by the geometry of the cylinder and crank mechanism. The new technology called variable compression ratio, and marked with the VCR symbol, completely changes the previously unchanged feature of the engines. Many companies have noticed significant potential hidden in variable compression systems and focused their entire attention on finding ways to apply and use this feature. Therefore, over the last years, there has been a growing interest in this type of innovative construction.

1.1 Potential application area of VCR technology

The origination of the idea of variable compression ratio was the desire to use for the engine supplying fuels with different properties; it is the so-called flex-fuel capability. Although it is still a valuable property, currently in the development of internal combustion engines, a special emphasis is placed on reducing fuel consumption, and thus CO_2 emissions, as well on reducing the emission of harmful exhaust components—carbon monoxide, hydrocarbons, nitrogen oxides, or particulates.

Global automotive industry as well science and research institutions involved in the development of VCR technology say about the potential for reducing fuel consumption by using a variable compression ratio in engines of different capacity. Significant economic benefits are particularly visible when variable compression ratio technology is used for high-power engines. It is significant also if these are naturally aspirated or boosted engines with direct injection (DFI) and variable-controlled valve actuation (VVA). **Figure 1** illustrates the potential benefits for the individual and combined use of these solutions.

Significantly, greater benefits in the field of fuel economy and engine ecology can be obtained by combining the VCR and VVA variable valve actuation systems (**Figure 2**). The use of these systems allows, for example, the implementation of the Atkinson working cycle and the using engines of smaller dimensions with maintained high operational parameters, that is, according to the idea of "ultra-downsizing."

"Downsizing" is a relatively new development trend of piston engines, which gives a response to strong pressure on reducing fuel consumption and improving the overall engine efficiency. The idea of downsizing is to reduce the cubic capacity of the powerdrives and increase the power ratio, mainly by using still higher and

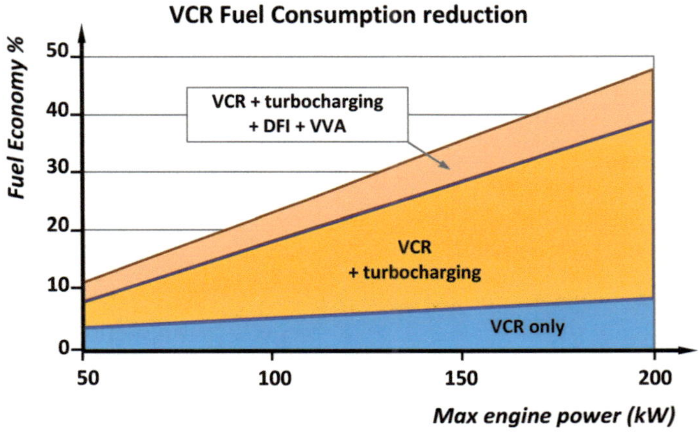

Figure 1.
The predicted increase in engine fuel economy with using VCR technology [1].

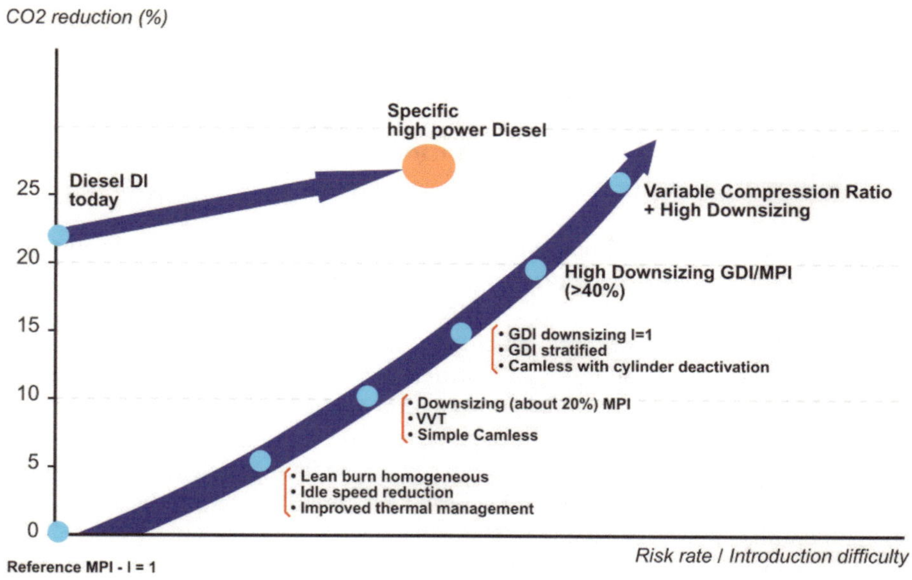

Figure 2.
The foreseen decrease of CO_2 emissions, thanks to the use of VCR and VVA technologies [1].

higher boosting pressure. Reduced stroke volume can be also obtained, among others, by reducing the number of cylinders. It means the mechanical and thermal losses are reduced, as well cylinder charge exchange losses, the so-called "pumping" losses too. At the same, the overall efficiency of the engine increases. On the other hand, an increase in the boost pressure requires a reduction of geometric compression ratio to avoid adverse phenomena in the combustion process. Due to the direct relationship with the cycle efficiency, it is preferred to use as high compression ratio as possible. So, the compression ratio value must be a compromise between achieving the high thermal efficiency of the cycle and other restrictions, such as: knock limit, level of mechanical and/or thermal loads, maximum rate of pressure rising (engine run hardness), etc. However, these limitations are only in force at high engine loads, especially at high boosting. In conventional engines, the compression ratio is optimized and finally determined for these operating conditions. However, this is not a significant area of their usual operation. At low and moderate loads, the compression ratio could be much higher, giving increased operational efficiency of the vehicle. Therefore, the VCR technology enables a significant extension of the "downsizing" range, allowing further reduction of the displacement volume and the use of even higher boosting pressures. It is estimated that this method of adjusting the working parameters of the spark-ignition engine can result in a reduction in the fuel consumption by up to 30% without a significant increase of toxic exhaust compound emission [1].

Finally, the next area of using VCR technology is the possibility of effective implementation of advanced, low-temperature LTC combustion processes (low-temperature combustion). They are characterized by extremely low emission levels of toxic compounds as a result of high complementarity of the combustion phase, but conducted under conditions with reduced charge temperature. Especially, a lack of temperature gradients in the combustion chamber is the main cause of avoiding nitrogen oxides' formation. There are, however, certain difficulties in obtaining low-temperature combustion stability under the high engine loads. Within this range of engine operation, a new VCR variable compression ratio technology exhibits a promising application potential, thanks to the high ability in moderating the thermodynamic conditions in the combustion chamber, which determine the initiation and course of combustion process.

2. Technical methods for implementation of variable compression ratio technology in piston engines

Theoretically, there are several possible methods of using the variable compression ratio VCR technology in piston engines. Some of them were used in prototype engines and they undergo operational tests.

Table 1 schematically presents selected VCR layouts together with a brief analysis of their technical and operational features, including advantages or disadvantages in relation to the conventional construction of the engines. Noteworthy is the solution (f) of a complex lever-gear crank system, developed and applied by the French research group MCE-5 Development [2], as well as the SAAB SVC engine [3, 4], according the principle (a) and solution (c) implemented in the FEV Motorentechnik research engine [5].

There are many specific and unique constructions of VCR engines or even engine ideas and patents. **Table 1** collects the best-known approaches for VCR engine:

a. articulated monohead—the SAAB completely functional engine SVC [3, 4],

b. piston of variable deck height—different layouts presented by Daimler-Benz and Ford, as well [6],

c. eccentrics on crankshaft bearings developed by FEV [5],

d. multilink rod-crank mechanisms developed by Nissan [7],

e. secondary moving piston or valve in cylinder head—different Ford and Volvo/Alvar proposals [6],

f. gear-based crank mechanisms by MCE-5 [1, 2],

g. precisely shifted cylinder block—cylinder head assembly—used in authors' own project [8, 9].

Each of above are presented and widely discussed by Shaik et al. [6]. The SAAB's SVC engine according to the solution (a) has been developed earlier by Larsen [3]. The compression ratio is variable from 8:1 to 14:1. Similarly to the shifted cylinder head method (g), it reveals good compression ratio control ability, but with slight change in piston kinematics. As a common drawback of both systems, a worse reliability and durability characteristic can be pointed. The solution (b) based on piston deck height variation uses a complicated special piston construction [6]. It also does not provide easy and precise control of compression ratio. Eccentric on main bearings (c) seems to be devoid of substantial disadvantages, but it makes the crankshaft block more complex. This solution is developed by FEV and used in their concept VCR car [5]. Nissan Motors developed a multilink rod-crank mechanism [6] according to the layout (d). It provides moderate compression ratio control ability at significant change in piston kinematics. Changing compression ratio using a small chamber with moving piston/valve (e) is relatively simple method to be applied in standard engines conversion into the VCR engines. As the drawback of this manner for changing compression ratio, the poor combustion chamber

DOI: http://dx.doi.org/10.5772/intechopen.93572

Design layouts	(a)	(b)	(c)	(d)	(e)	(f)	(g)
Technical and operational features	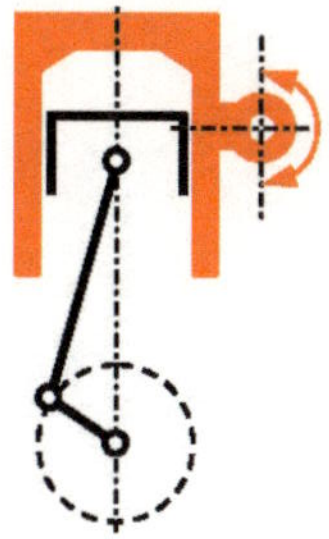 [3, 4]	 [6]	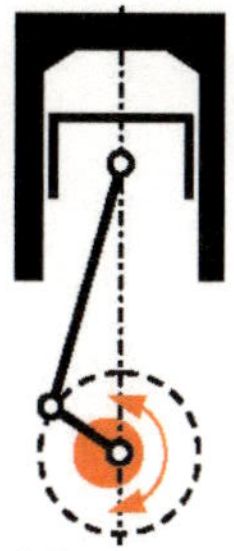 [5]	 [7]	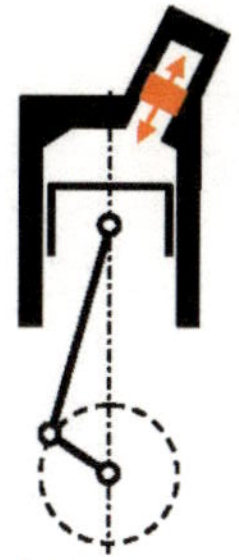 [6]	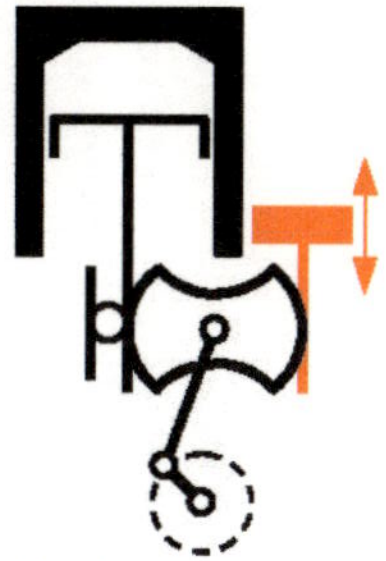 [1, 2]	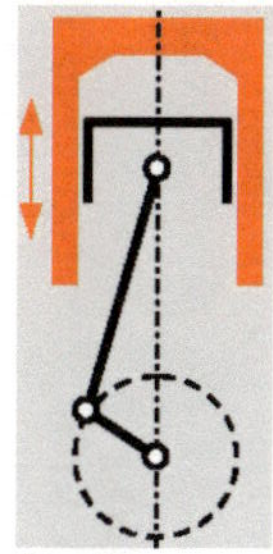 [8, 9]
Combustion chamber consistency	Neutral	Neutral	Neutral	Neutral	↘↘↘	Neutral	Neutral
Cranckshaft mechanism kinematic	≠	Neutral	≠	≠≠≠	Neutral	≠≠≠	Neutral
Mechanical and friction losses	Neutral	↘	Neutral	↘↘	Neutral	↗↗	Neutral
Engine design integrity and dimensions	↘	↘	↘	↘	↘	↘	↘
Impact of variability of compression ratio on cyl. capacity	Neutral	Neutral	Neutral	↘↘	Neutral	Neutral	Neutral
Accuracy of control and range of changes in compression ratio	↑↑	↓↓↓	↑↑	↑	↑	↑↑	↑↑↑

↗—improvement, ↘—deterioration, ↑—high mark, ↓—low mark, and ≠—distinctive mark.

Table 1.
Technical methods of application VCR technology in piston engines and their technical and operational features.

integrity can be pointed [6]. The gear-based crank mechanism (f) is very advanced technique extensively developed by MCE-5 research group [1, 2]. It shows high precision in CR control and profitable changes in piston kinematics that avoids side forces acting on the piston.

Analyzing the possible solutions of VCR engines, both hypothetical constructions and actual prototype units, two general strategies for changing the value of the compression ratio during the engine run can be noticed.

1. The change and control of the compression ratio at the assumed level takes place by changing the position or geometry of the engine part, which consist of the cylinder head assembly. This method does not interfere with the moving parts of the crank-piston system, thus the friction losses and kinematics of the crank-piston system during engine operation remain unchanged or change in a very small extent comparing to a conventional engine.

2. The change of the compression ratio is a result of the intended changes in the geometry and/or kinematics of the crank-piston system due to special constructions of mechanisms allowing for the correction of the distance between the top plane of the piston and the bottom plane of the head. In this case the power of friction losses in the crank-piston system usually increases, although it is also possible to be reduced (e.g., solution (f)—**Table 1**). Sometimes variation of the compression ratio according to these concepts also causes an unfavorable change in the cylinder stroke volume (e.g., solution (d)—**Table 1**).

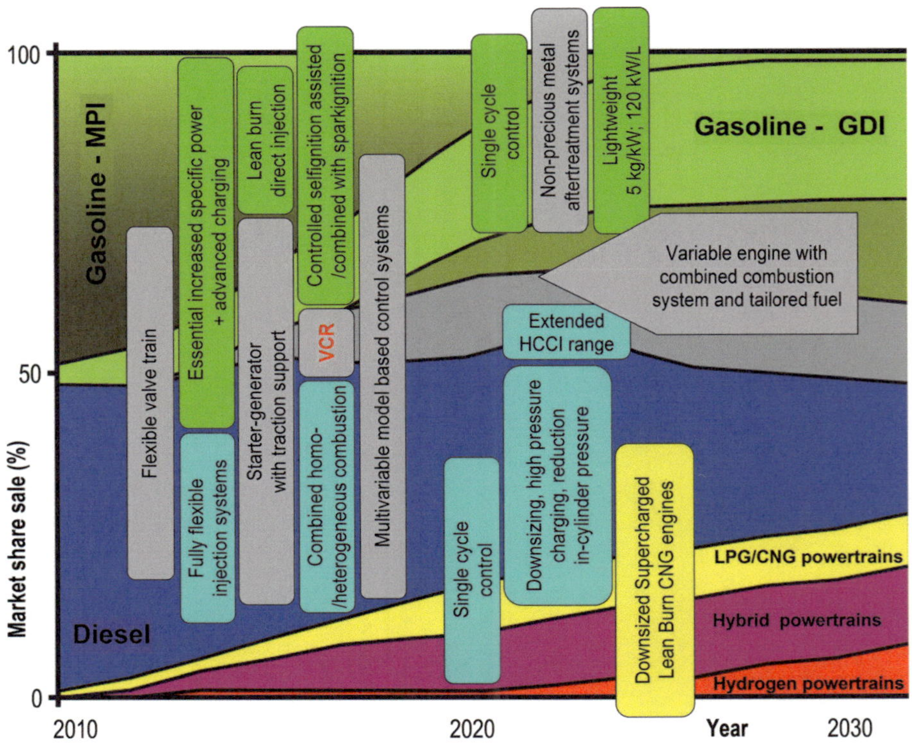

Figure 3.
Forecasts for the development and share of selected advanced technologies in the combustion powerdrives of motor vehicles [10].

Despite the fact that the idea of variable-pressure VCR engines is associated with significant construction and technological complexity and many operational problems, it is estimated that it starts to become a technical standard for automotive piston engines in the near future (see **Figure 3**).

Therefore, it was accepted that development and building a prototype variable compression ratio engine, along with the acquired knowledge and experience, will allow for successive improvement of the design and, as a result, get a fully functional, unique research instrument that makes possible to perform a number of innovative scientific works.

3. Design assumptions for the test engine and selection of the base unit

On the basis of the analyses in the field of variable compression ratio technology, taking into account all advantages and disadvantages of the known technologies for of variable compression ratio engines and existing prototypes as well as own manufacturing capabilities, it was decided that the construction of the test engine will be carried out according to for the kinematic system shown on case (g) in **Table 1**, that is, consisting of controlled positioning and movement of the cylinder head assembly along the cylinder axis. This solution is characterized by relatively low implementation costs due to the possibility of conversion of a standard piston engine into the VCR one, simplicity of construction and control, while ensuring a relatively wide range of changes of compression ratio at a high accuracy in positioning. The engine will be based on a serial produced combustion engine. It was assumed that for the purposes of the assumed scope of experimental tests, it will be necessary to obtain a wide range of compression ratio variation covering typical values for both spark and diesel engines, that is, from around 9:1 up to 19:1. The test engine should be also liquid cooled to ensure good temperature stability during research.

A medium-speed, liquid-cooled 4-cylinder diesel engine manufactured by VEB IFA-Motorenwerk Nordhausen type 4 VD 14.5/12-1 SRW was selected for the

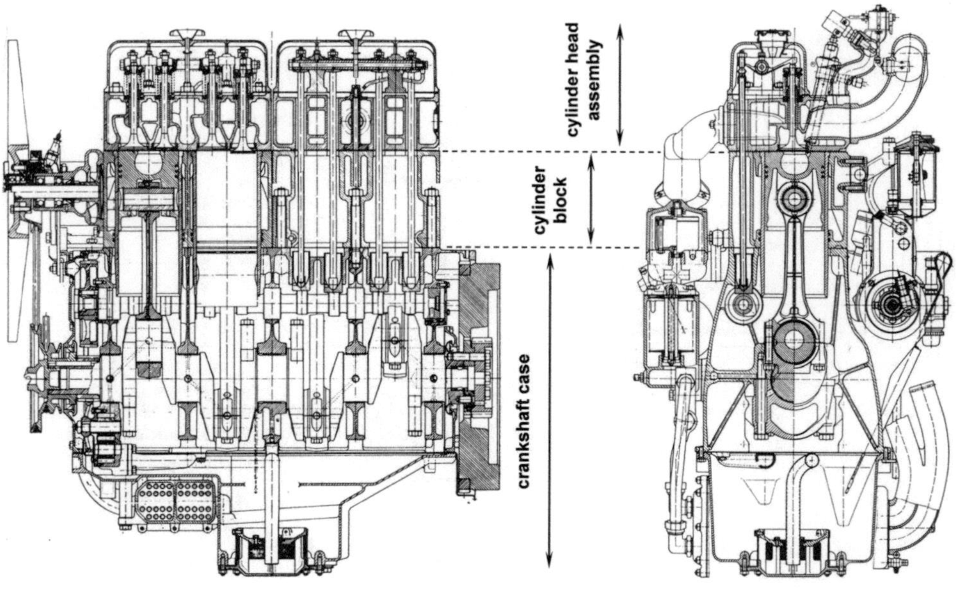

Figure 4.
The main cross sections of the 4 VD 14.5/12-1 SRW engine as the basis for the own designed research unit in the VCR technology with the dividing planes of the engine body shown [11].

construction of the VCR engine. The unique structural feature of this engine, which decided on its selection, was the physically existing plane that divides the crankcase from the cylinder assembly (see **Figure 4**).

The basic parameters of the 4 VD engine are as follows: cylinder diameter—120 mm, piston stroke—145 mm, displacement—6560 ccm, original geometric compression ratio—18:1, valve drive system: OHV overhead valves with camshaft located in the crankshaft block, cam followers, sticks, valve arms mounted on the axle above the head. Detailed engine specifications are described in **Table 2**.

Geometric dimensions of the 4 VD 14.5/12-1 SRW engine allowed to determine how the compression ratio will change with the cylinder head assembly moving along the cylinder axis according to the relationship (Eq. (1)):

$$\varepsilon' = \frac{\frac{\varepsilon \cdot V_c}{\varepsilon - 1} + \pi \frac{D^2}{4} \cdot h}{\frac{V_c}{\varepsilon - 1} + \pi \frac{D^2}{4} \cdot h} : 1 \qquad (1)$$

where ε' is compression ratio as a function of the cylinder head assembly shift, ε is original compression ratio, V_c is cylinder displacement, D is cylinder diameter, h is shift value of cylinder head assembly relative to the initial position.

Eq. (1), substituted with appropriate values, shows that the range of the cylinder head assembly tilting from initial position up to 10 mm travel gives the compression ratio changes from 19:1 to 9:1, according to the curve shown in the **Figure 5**. The

Producer	**VEB IFA-Motorenwerk Nordhausen**
Type	4-stroke, diesel
Number of cylinders	4
Ignition order	1-3-4-2
Cylinder layout	In-line
Piston stroke	145 mm
Cylinder diameter	120 mm
Displacement	6560 ccm
Compression ratio	18:1
Rated power	92 kW (125 KM)
Cranckshaft speed at rated power	2300 rpm
Maximum torque	430 Nm
Cranckshaft speed at max. torque	1350 rpm
Mean effective pressure	0.77 MPa
Specific fuel consumption for rated power	240 g/kW·h (175 g/KM·h)
Minimum specific fuel consumption	218 g/kW·h (160 g/KM·h)
Lubrication system	Closed circulation, pressurized
Fuel delivery	Direct injection (system MAN), single-hole sprayer, in-line section fuel pump with mechanical regulation of engine speed
Initial pressure of injector opening	17.5 MPa
Engine starter	Electric motor
Power and supply voltage of engine starter	3 kW, 24 V

Table 2.
Original technical specification of the 4 VD 14.5/12-1 SRW engine [11].

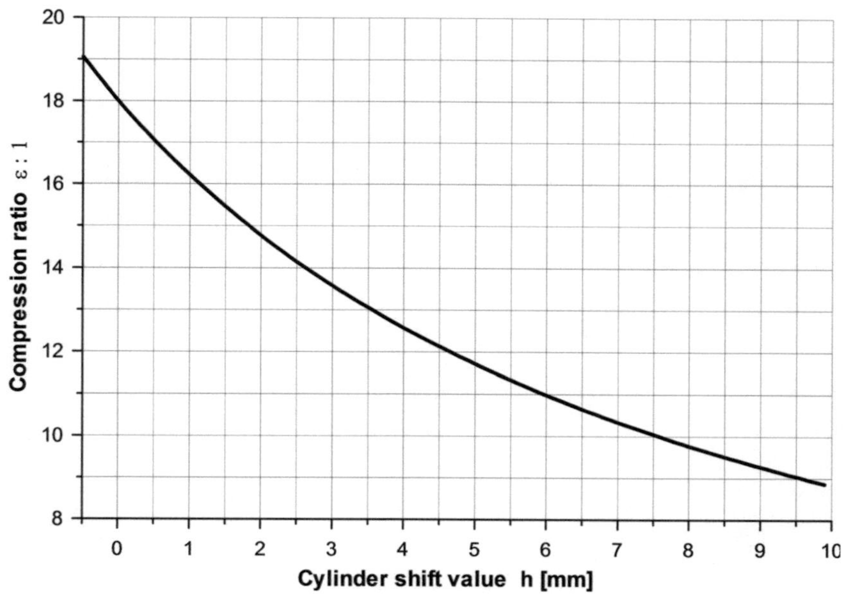

Figure 5.
Compression ratio versus cylinder shift value for 4 VD 14.5/12-1 SRW engine.

range of these changes essentially coincides with the desired research scope of the engine. If necessary, it can be changed relatively easily by replacing the pistons with different volume of the combustion chamber [8, 9].

The adopted concept of changing the compression ratio requires the use of an accurate, precise cylinder head assembly shifting mechanism in relation to the crankshaft block. The requirement of this mechanism, in addition to the high accuracy of positioning and rapid change of the position of the cylinders, is to transfer gaseous forces generated by the combustion process in individual engine cylinders. The value of these forces can be determined from Eq. (2):

$$F_{max} = p_{max} \cdot \frac{\pi \cdot D^2}{4} \tag{2}$$

where F_{max} is the maximum force acting on the shifting system due to the gas pressure in the cylinder, p_{max} is maximum gas pressure in the cylinder, and D is cylinder diameter.

Assuming the maximum gas pressure in cylinders at 10 MPa, we obtain the force generated by a single cylinder at the level of 113 kN. Hence, the sliding mechanism must have adequate strength, but also rigidity, operational reliability, small dimensions and relatively high positioning resolution, especially in the range of high values of compression ratio.

The task of the newly designed mechanism for compression ratio changing is precise shifting the 4-cylinder "cylinder head" assembly in the range of 0–10 mm by means of synchronously rotating two eccentric shafts that are connected to the sliding elements by kind of yokes—connecting rods. The layout diagram and the source of the main mechanical loads are shown in **Figure 6**. The analysis of forces and torques shows that the eccentric shafts will be loaded with a twisting torque Ms of approx. 300 Nm each.

Figure 7 shows the location of eccentric shafts together with the shaft drive system. Both shafts have bearings on both ends and in the middle of their length. At the free ends of both shafts there are two geared right-angle power transmissions; their inputs are connected by a common drive shaft that is driven by a synchronous

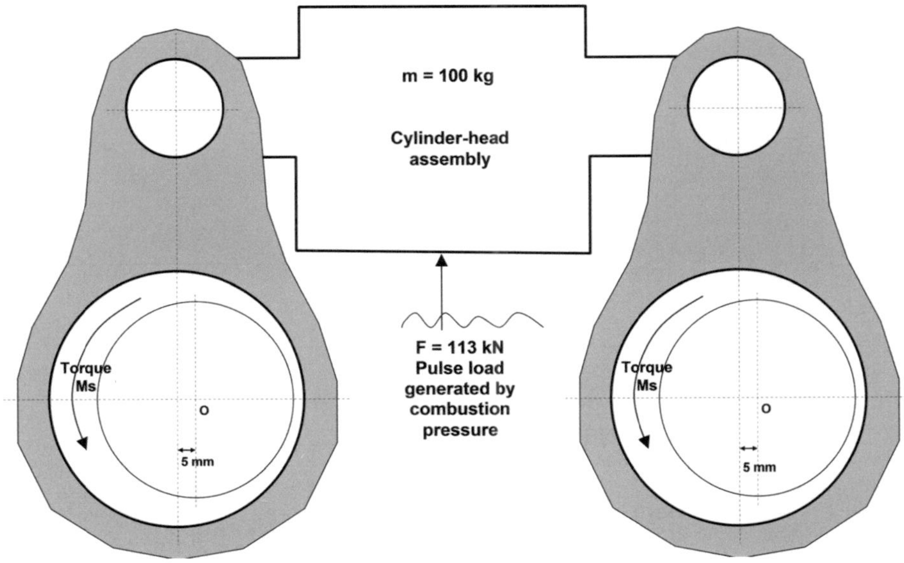

Figure 6.
Diagram of the yoke-eccentric cylinder sliding mechanism and the main sources of mechanical loads.

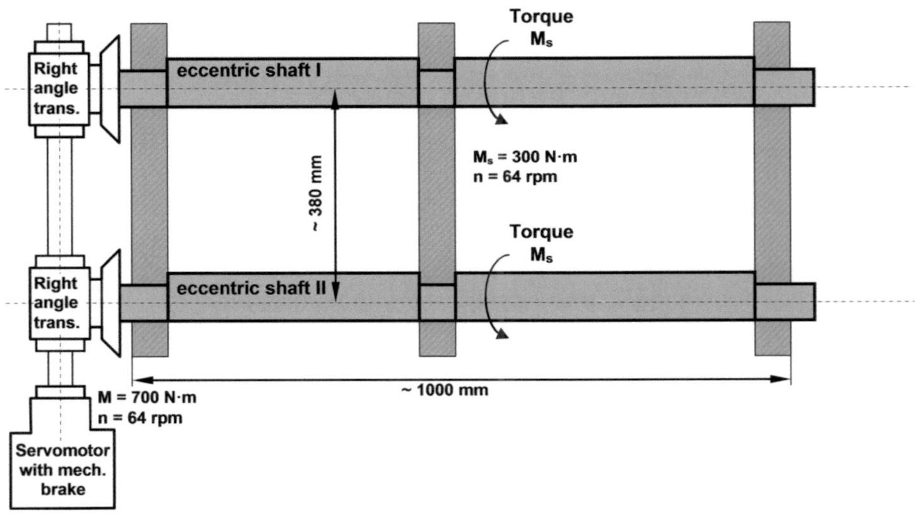

Figure 7.
Project of the location of eccentric shafts with the drive system diagram.

servomotor with a mechanical brake. The wheelbase of the shafts is approx. 380 mm, which directly fits to the design features of the test engine.

Analysis of loads shows that each of the bevel-geared transmissions has to transfer a maximum torque less than 300 Nm, while the main servomotor has to generate a double value of that torque, that is, at least 600 Nm.

Therefore, an appropriate selection of working elements has been made, taking into account the structural safety factors. The bevel gearboxes TRAMEC in the version RA 38AC 1:1 E B3 with the rated torque Ma = 320 Nm are used as the right-angle drives. As the servomotor a two-stage flat reducer, type Stöber SMS version F402AGN0470 EZ503U EL1, with an acceleration torque of 700 Nm, driven by the Stöber POSIDRIVE MDS5110A/L 11.0 kW 3 × 400 V inverter is used (**Figure 8**). All gear units are with reduced mid-gear lash to the value below 10 arc minutes.

Figure 8.
Overall views of the RA type TRAMEC bevel gearbox, Stöber SMS servomotor, and Stöber POSIDRIVE MDS inverter.

Construction work began with the reverse engineering process on the research engine, that is, scanning the spatial engine body with the determination of characteristic points, surfaces, distances, clearances, etc., allowing the design and manufacture of other engine components and subassemblies. As previously mentioned, the characteristic feature of IFA 4 VD 14.5/12-1 SRW engine, which influenced the decision on its selection for adaptation works is the fact that the engine block is not permanently fixed to the cylinders. The division plane is shown in **Figure 9**. Its physical presence gave the opportunity to design and build an appropriate drive mechanism to realize the movement of the cylinders (together with the heads) in the vertical direction.

The basic elements of the cylinder sliding system are the servomotors chosen with catalogs, two bevel gears, drive couplings, and two eccentric shafts, connecting rods and a designed and made main support plate that keeps the

Figure 9.
3D scanning image of the IFA 4 VD engine as a base for performing necessary design and technological changes; the green plane is marked showing the division plane between the crankcase block and the cylinder blocks.

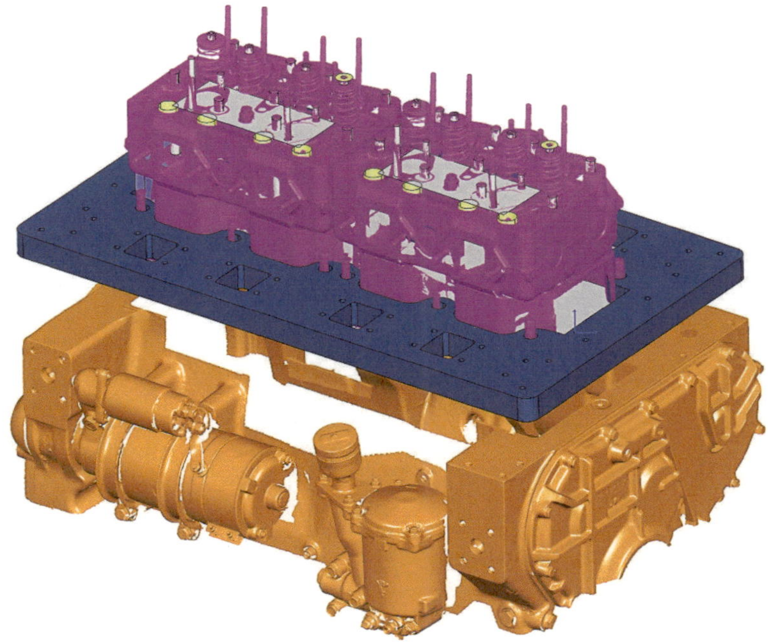

Figure 10.
Design of the cylinder support plate and its mounting on the screws fitted into the cylinder blocks.

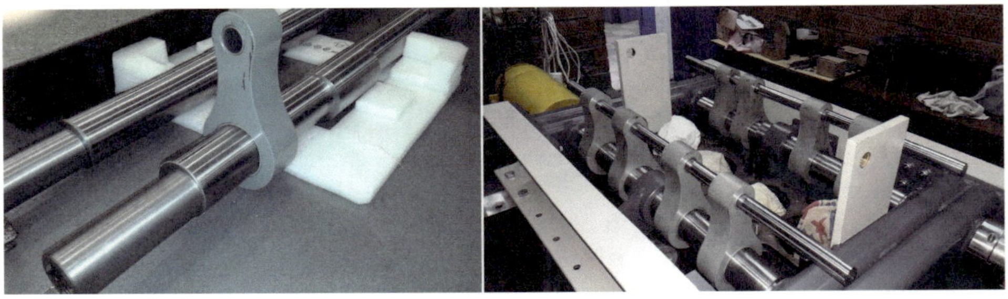

Figure 11.
Connecting yokes mounted with the eccentric shafts.

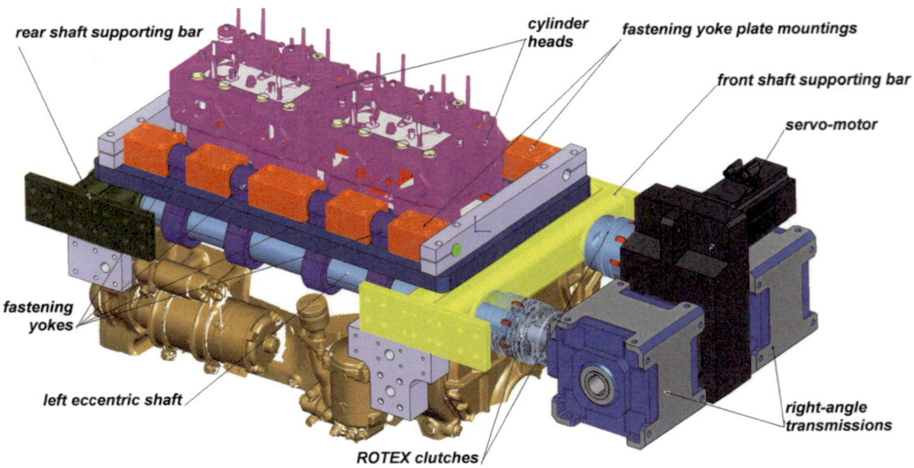

Figure 12.
View of the assembly of the cylinder blocks sliding system.

Figure 13.
An isometric view of the VCR engine with assembled with the cylinder sliding system and the drive mechanism of eccentric shafts.

cylinders. The main board (**Figure 10**, dark blue color) is seated and attached to the cylinder blocks using special threaded mounting bolts.

To allow the additional space for insertion of a flexible seal gasket between crankshaft block and cylinder blocks it was necessary to change the pistons with the larger height ones by at least 5 mm. Suitable pistons that meet the necessary

Figure 14.
View of the complete VCR engine during functional tests.

dimensions for the designed VCR engine were adopted from the HANOMAG D942 engines.

The next key stage in the VCR engine design process was the exact machining of eccentric shafts for the cylinder sliding system together with a set of eight connecting yokes. To ensure proper dimensional accuracy, technological work was carried out on CNC machine tools. **Figure 11** shows the connecting yokes fastened with the eccentric shafts.

The movement of the supporting plate together with the cylinders is enforced by an eccentric-crank mechanism driven by a complex servomechanism (**Figure 12**), that is, through two angle bevel gears and two eccentric shafts. The elements shown in orange make the assembly fixing the axle passing through the connecting rod holes. Eccentric shafts are mounted in sleeves welded to the transverse beams (green and yellow). The drive for eccentric shafts is transmitted via two ROTEX GS clutches. Any disorders and risk of misalignments in movement of the cylinder-cylinder head assembly is secured by set of sliding barrels. Sliding barrels are permanently fixed in the crankcase and fitted into the precise holes made in the cylinders body.

The entire design of the VCR engine, for a better illustration of its structure complexity, is presented in isometric views in **Figure 13**, while the finally completed research engine with variable compression ratio is presented in **Figure 14**.

4. Conclusion

Piston engine for many decades is a basic and commonly used source of mechanical drives in various types of machinery and technical equipment, including motor vehicles and other means of transportation. Despite the various controversial forecasts and views that have recently appeared, and not always are based on trusted and documented technical knowledge, the combustion engine will certainly remain an irreplaceable source of propulsion for many branches of transport and industry. One should keep in mind the intense scientific, technical, and technological progress that makes the final product even more and more technically perfect. Taking into account the current development trends that arisen from an experience of recent years, which are focused mainly on improving combustion processes, it can be noted that the presented technical development of VCR internal combustion engines gives a significant contribution in this process.

A great innovation and application potential is shown by the worked out original design, constructional and technological achievements covering a four-cylinder combustion engine with a variable compression ratio VCR feature. Attempts to develop such an original powerdrive unit, except for some major automotive industry efforts, were usually finished unsuccessfully. The developed design of the VCR engine opens up new research and development opportunities that were not available before. It concerns mainly to new directions of improvement of engine working processes and exploitation possibilities of internal combustion engines, that is, research on advanced, low-temperature combustion processes or research on the unification and flexible use of various fuels for transportation, including alternative fuels of different reactivities.

Acknowledgements

Authors wish to thank to the Ministry for Science and Higher Education in the Republic of Poland for their financial support of this work.

Author details

Paweł Woś*, Krzysztof Balawender, Mirosław Jakubowski, Artur Jaworski, Paulina Szymczuk and Adam Ustrzycki
The Faculty of Mechanical Engineering and Aeronautics, Rzeszów University of Technology, Rzeszów, Poland

*Address all correspondence to: pwos@prz.edu.pl

References

[1] MCE-5 VCRi: Pushing back the fuel consumption reduction limits, MCE-5 Development 2010 [Internet]. Available from: www.vcri.eu [Accessed: 5 October 2018]

[2] Rabhi V, Beroff J, Dionnet F. Study of a Gear-Based Variable Compression Ratio Engine. SAE Paper No. 2004-01-2931. Warrendale PA, USA: SAE International; 2004

[3] Larsen GJ. Research Engine for Evaluating the Effects of Variable Compression Ratio (VCR) and/or Variable Valve Timing (VVT). SAE Paper No. 910053. Warrendale PA, USA: SAE International; 1991

[4] SVC Variable Compression Engine, Linköping University, Sweden 2014 [Internet]. Available from: www.vehicular.isy.liu.se [Accessed: 5 October 2018]

[5] Habermann K. Demonstration Vehicle with Continuously Variable Compression Ratio (VCR). FEV Magazine “Spectrum”, Issue 22, 2003 [Internet]. Available from: http://www.fev.com [Accessed: 5 October 2018]

[6] Shaik A, Shenbaga Vinayaga Moorthi N, Rudramoorthy R. Variable compression ratio engine: A future power plant for automobiles—An overview. Proceedings of the Institution of Mechanical Engineers (IMechE); Part D: Journal of Automobile Engineering. 2007;**221**(D9):1159-1168

[7] Moteki K, Aoyama S, Ushijima K, Hiyoshi R, Takemura S, Fujimoto H, et al. A study of a variable compression ratio system with a multi-link mechanism. SAE Paper No. 2003-01-0921. Warrendale PA, USA: SAE International; 2003

[8] Woś P, Balawender K, Jakubowski M, Kuszewski H, Lejda K, Ustrzycki A. Design of Affordable Multi-Cylinder Variable Compression Ratio (VCR) Engine for Advanced Combustion Research Purposes. SAE Paper No. 2012-01-0414. Warrendale PA, USA: SAE International; 2012

[9] Woś P, Jakubowski M. Variable Compression Ratio Engine. Patent No. 217826. Warsaw: The Patent Office of the Republic of Poland; 29 August 2014

[10] EUCAR R&D for Europe, The European Council for Automotive R&D Conference [Internet]. 2006. Available from: www.eucar.be [Accessed: 5 October 2018]

[11] VEB IFA-Automobilwerke Ludwigsfelde (ed.), Service Manual for LKW IFA W50. VEB Fachbuchverlag, Leipzig 1972 (in German)

Chapter 7

Stability Analysis of Long Combination Vehicles Using Davies Method

Gonzalo Guillermo Moreno Contreras,
Rodrigo de Souza Vieira and Daniel Martins

Abstract

The cargo transportation in the world is mostly dominated by road transport, using long combination vehicles (*LCV's*). These vehicles offer more load capacity, which reduces transport costs and thus increases the efficiency and competitiveness of companies and the country. But the tradeoff of *LCV's* is their low lateral stability and propensity to roll over, which has been the focus of many studies. Most vehicle stability models do not consider the longitudinal aspects of the vehicle and the road, such as the stiffness of the chassis, the gravity center location, and the longitudinal slope angle of the road. But, the use of three-dimensional models of vehicles allows a more rigorous analysis of vehicle stability. In this context, this study aims to develop a three-dimensional mechanism model representing the last trailer unit of an *LCV* under an increasing lateral load until it reaches the rollover threshold. The proposed model considers the gravity center movement of the trailer, which is affected by the suspension, tires, fifth-wheel, and the chassis. Davies method has proved to be an important tool in the kinetostatic analysis of mechanisms, and therefore it is employed for the kinetostatic analysis of the three-dimensional mechanism of the trailer.

Keywords: stability, road safety, static rollover threshold (SRT)

1. Vehicle model for lateral stability

According to Rempel [1] and Melo [2], the last unit (semi-trailer) of an *LCV* is the critical unit, since it is subjected to a high lateral acceleration compared to the tractor unit, which impacts the rollover threshold of the unit and the vehicle. Taking into account this aspect, a simplified trailer model (**Figure 1**) is modelled and analysed to calculate the *SRT* factor for *LVCs*.

The tyres, suspension, fifth wheel, and chassis are directly responsible for the *CG* movements; these movements are dependent on the forces acting on the trailer *CG*, such as weight (*W*), disturbance forces imposed by the ground, and lateral inertial force (ma_y) when the vehicle makes a turn. During cornering or evasive manoeuvres, the weight and the lateral inertial force acting on the vehicle centre of gravity cause its displacement, which can lead to vehicle rollover.

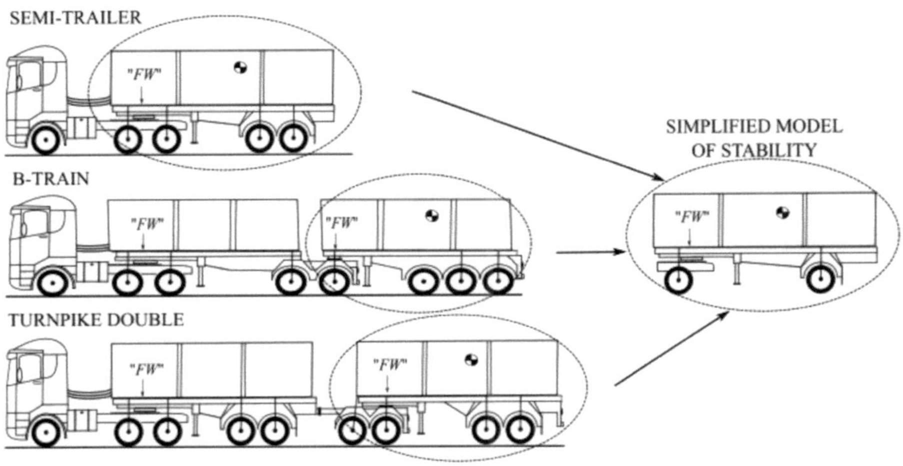

Figure 1.
Simplified trailer model.

1.1 Tyres system

The tyres system (tyres and rigid suspension) maintains contact with the ground and filters the disturbances imposed by road imperfections [3]. This system allows two motions of the vehicle: displacement in the z-direction and a roll rotation around the x-axis [4], as shown in **Figure 2**.

1.1.1 *Kinematic chain for tyres system*

Mechanical systems can be represented by kinematic chains composed of links and joints, which facilitate their modelling and analysis [5–7].

The kinematic chain of the tyres system in **Figure 2** has 2-DoF ($M = 2$), the workspace is planar ($\lambda = 3$), and the number of independent loops is one ($\nu = 1$). Based on the mobility equation, the kinematic chain of tyres system should be composed of five links ($n = 5$) and five joints ($j = 5$) [7].

To model this system, the following considerations were taken into account:

- There are up to three different components of forces acting on the tyre-ground contact i of the vehicle [8–10], as shown in **Figure 3**, where F_{xi} is the traction or brake force, F_{yi} is the lateral force, and F_{zi} is the normal force;

- However, at rollover threshold, tyres 1 and 4 (outer tyres in the turn, **Figure 4**) receive greater normal force than tyres 2 and 3 (inner tyre in the turn, **Figure 4**), and thus tyres 1 and 4 are not prone to slide laterally. We consider that tyres 1 and 4 only allow vehicle rotation along the x-axis. Therefore, tyre-ground contact was modelled as a pure revolute joint R along x-axis.

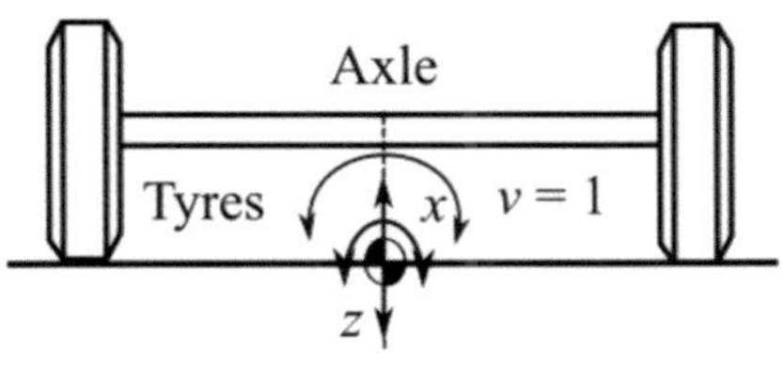

Figure 2.
Tyres system.

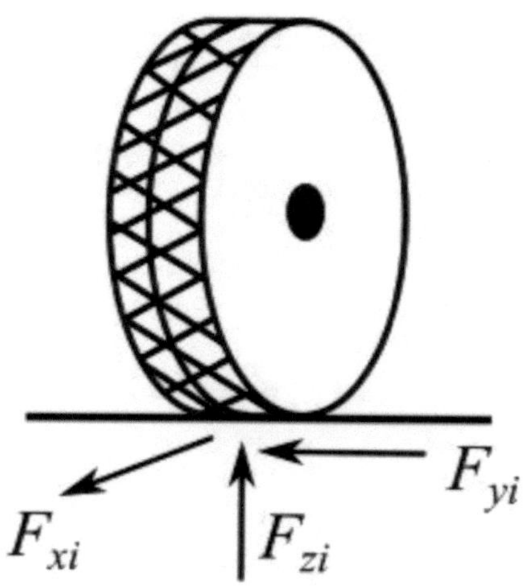

Figure 3.
Movement constraints in Tyre-road contact.

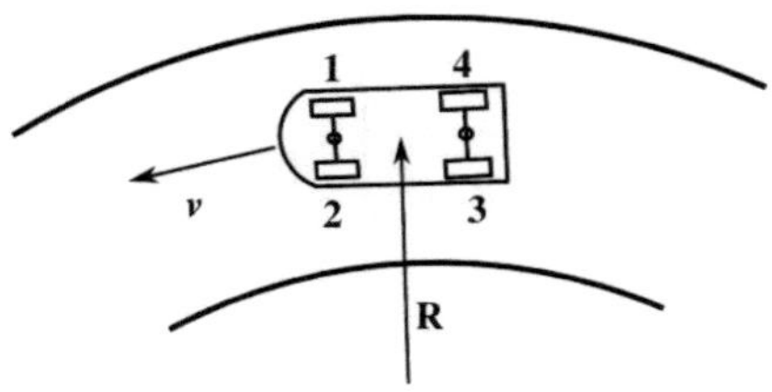

Figure 4.
Vehicle on a curved path.

- While tyres 2 and 3 have a lateral deformation and may slide laterally, producing a track width change of their respective axles. As a consequence, tyres 2 and 3 have only a constraint on the *z*-direction. Therefore, tyre-ground contact was modelled as a prismatic joint *P* in the *y*-direction.

- Tyres are assumed as flexible mechanical components and can be represented by prismatic joints *P*, [11, 12].

- In vehicles with rigid suspension, tyres remain perpendicular to the axle all the time.

Applying these constraints, **Figure 5a** shows the proposed kinematic chain model of the tyres system.

The kinematic chain is composed of five links identified by letters A (road), B (outer tyre in the turn), C and D (inner tyre in the turn), and E (vehicle axle); and the five joints are identified by numbers as follows: two revolute joints *R* (tyre-road

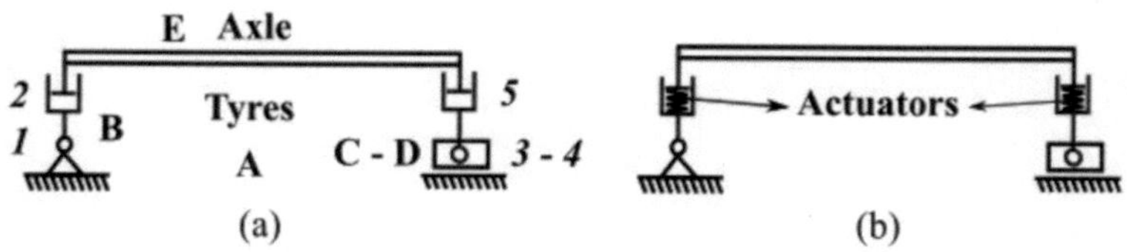

Figure 5.
(a) Kinematic chain of the tyres system. (b) Tyres system including actuators.

contact of joints 1 and 4) and three prismatic joints P, two that represent tyres of the system (2 and 5), and one the lateral slide of tyre 2 (3).

The mechanism of **Figure 5a** has 2-DoF, and it requires two actuators to control its movement. The mechanism has a passive actuator in each prismatic joint of tyres (2 and 5—axial deformation); these actuators control the movement along the x- and z-axes, as shown in **Figure 5b**.

In this model, the revolute joint (3) and the prismatic joint (4) can be changed by a spherical slider joint (S_d), with constraint in the z-axis, as shown in **Figure 6**.

1.1.2 Kinematics of tyre system

The movement of this system is orientated by the forces acting on the mechanism (trailer weight (W) and the inertial force (ma_y)) [13]. These forces affect the passive actuators of the mechanism, as shown in **Figure 7**.

Eqs. (1)–(5) define the kinematics of the tyres system.

$$l_i = \delta_T + l_r = \frac{3\Delta F}{k_t + a_c} + l_r \approx \frac{-F_{Ti} + F_{zi}^{start}}{k_T} + l_r \tag{1}$$

$$\beta_i = 90 - arcsin\left(\frac{l_{i+1}}{\sqrt{t_{i+1}^2 + l_{i+1}^2}}\right) \tag{2}$$

$$t_i = \sqrt{t_{i+1}^2 + l_{i+1}^2 + l_i^2 - 2\left(\sqrt{t_{i+1}^2 + l_{i+1}^2}\right) l_i\ cos\,(\beta_i)} \tag{3}$$

$$\delta_i = arcsin\,(l_i\ sin\,(\beta_i)/t_i) \tag{4}$$

$$\theta_i = \theta_j = 90 - \delta_i - \beta_i \tag{5}$$

where δ_T is the normal deformation of the tyre [14], ΔF is the algebraic change in the initial load, k_t is the vertical stiffness of the tyre, a_c is the regression coefficient, F_{Ti} is the instantaneous tyre normal load, l_i is the instantaneous dynamic rolling radius of the tyre i, F_{zi}^{start} is the initial normal load i, k_T is the equivalent tyre vertical stiffness, l_r is the initial dynamic rolling radius of tyre i, t_i is the track width,

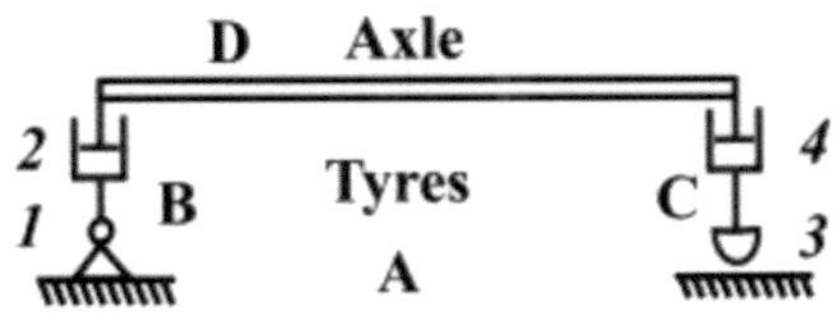

Figure 6.
Tyres system model.

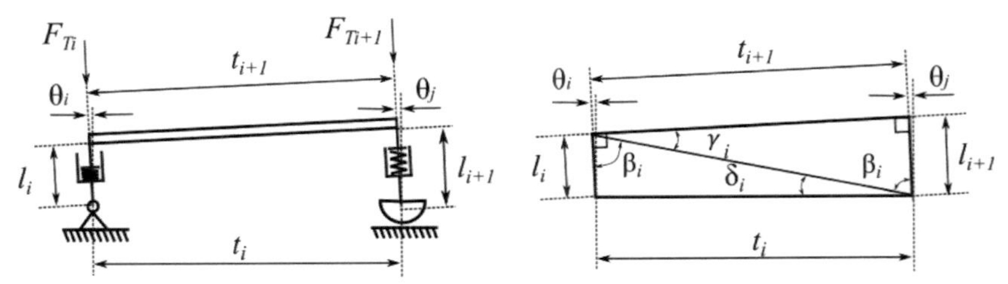

Figure 7.
Movement of tyres system.

t_{i+1} is the axle width, and $\theta_{i;j}$ are the rotation angles of the revolute joints i and j respectively.

1.2 Suspension system

This system comprises the linkage between the sprung and unsprung masses of a vehicle, which reduces the movement of the sprung mass, allowing tyres to maintain contact with the ground, and filtering disturbances imposed by the ground [3]. In heavy vehicles, the suspension system most used is the leaf spring suspension or rigid suspension [15], as shown in **Figure 8**. For developing this model (trailer), it is assumed that the vehicle has this suspension on the front and rear axles.

The rigid suspension is a mechanism that allows the following movements of the vehicle's body under the action of lateral forces: displacement in the z- and y-direction and a roll rotation about the x-axis [1, 8], as shown in **Figure 9a** and **b**.

1.2.1 Kinematic chain of the suspension system

The system of **Figure 10a** has 3-DoF (M = 3), the workspace is planar ($\lambda = 3$) and the number of independent loops is one ($\nu = 1$). From the mobility equation, the kinematic chain of suspension system should be composed of six links ($n = 6$) and six joints ($j = 6$).

To model this system the following consideration is considered: leaf springs are assumed as flexible mechanical components with an axial deformation and a small shear deformation, and can be represented by prismatic joints P supported in revolute joints R [16].

To allow the rotation of the body in the z-axis, the link between the body and the leaf spring is made with revolute joint. Applying these concepts to the system, a model with the configuration shown in **Figure 10b** is proposed.

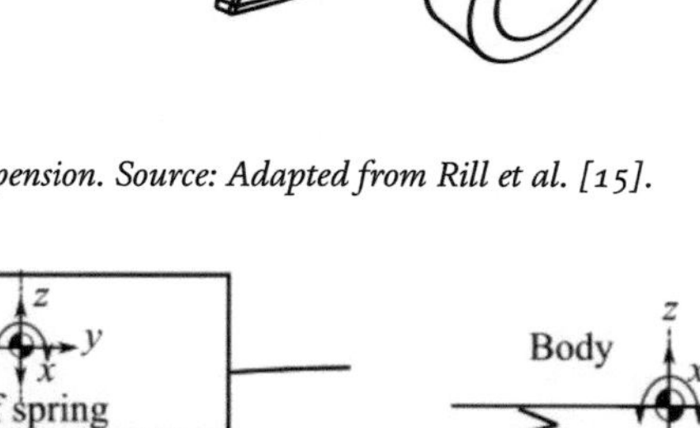

Figure 8.
Solid axle with leaf spring suspension. Source: Adapted from Rill et al. [15].

Figure 9.
(a) Body motion. (b) Suspension system.

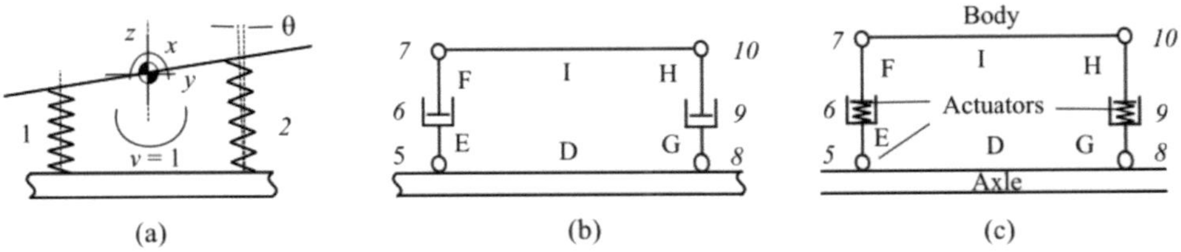

Figure 10.
(a) Movement of suspension system. (b) Kinematic chain of suspension system. (c) Suspension system including actuators.

The system is composed of six links identified by letters D (vehicle axle), E and F (spring 1), G and H (spring 2), and I (the vehicle body), and the six joints identified by the following numbers: four revolute joints R (5, 7, 8, and 10) and two prismatic joints P that represent the leaf springs of the system (6 and 9), as shown in **Figure 10b**.

The mechanism of **Figure 10b** has 3-DoF, and it requires three actuators to control its movements, applying the technique developed in Section 1.1, the kinematic chain has a passive actuator in the prismatic joints 6 and 9 (axial deformation of the leaf spring), and a passive actuator in the joints 6 and 9 (torsion spring—shear deformation); but the mechanism with four passive actuators is over-constrained, in this case only one equivalent passive actuator is used in the joint 5 or 8, as shown in **Figure 10c**.

1.2.2 Kinematics of the suspension system

The movement of the suspension is orientated first by the movement of the tyres system, and second by forces acting on the mechanism (vehicle weight (W) and inertial force (ma_y)). These forces affect the passive actuators of the mechanism, as shown in **Figure 11**.

Eqs. (6)–(12) define the kinematics of the suspension system:

$$\theta_n = {}^{T_{xn}}\!/\!_{k_{ts}} \tag{6}$$

$$l_n = \delta_{LS} + l_s = \frac{3\Delta F l^3}{8E_s NBT^3} + l_s \approx \frac{-F_{LSn} + F_{zi}^{start}}{k_{Ls}} + l_s \tag{7}$$

$$r = \sqrt{l_n^2 + b^2 - 2l_n b\ cos\,(90 + \theta_6)} \tag{8}$$

$$\beta_n = arccos\left((b^2 + r^2 - l_4^2)/(2br)\right) \tag{9}$$

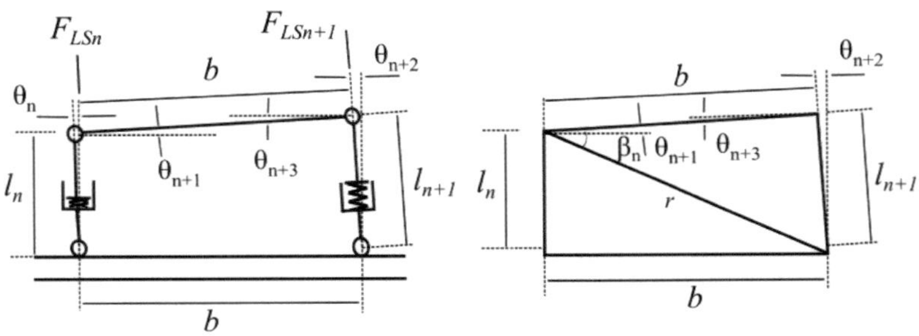

Figure 11.
Movement of suspension system.

$$\theta_{n+1} = \beta_n + arcsin\left(\frac{b}{r} sin\,(90 + \theta_n)\right) - 90 \tag{10}$$

$$\theta_{n+2} = \theta_n + arcsin\left(\frac{b}{r} sin\,(90 + \theta_n)\right) - arcsin\left(\frac{b}{l_{n+1}} sin\,(\beta_n)\right) \tag{11}$$

$$\theta_{n+3} = 90 - \beta_n - arcsin\left(\frac{b}{l_{n+1}} sin\,(\beta_n)\right) \tag{12}$$

where T_{xn} is the moment around the x-axis on the joint n, k_{ts} is the spring's torsion coefficient, δ_{LS} is the leaf spring deformation [17], ΔF is the algebraic change in the initial load, l is the length of the leaf spring, N is the number of leaves, B is the width of the leaf, T is the thickness of the leaf, E_s is the modulus of elasticity of a multiple leaf, l_n is the instantaneous height of the leaf spring n, F_{LSn} is the spring normal force n, l_s is the initial suspension height, k_{Ls} is the equivalent stiffness of the suspension, l_n is the instantaneous height of the leaf spring n, b is the lateral separation between the springs, and θ_n is the rotation angle of the revolute joint n.

1.3 The fifth wheel system

This system is a coupling device between the tractive unit and the trailer; but in the case of a multiple trailer train, a fifth wheel also can be located on a lead trailer. The fifth wheel allows articulation between the tractive and the towed units.

This system consists of a wheel-shaped deck plate usually designed to tilt or oscillate on mounting pins. The assembly is bolted to the frame of the tractive unit. A sector is cut away in the fifth wheel plate (sometimes called a throat), allowing a trailer kingpin to engage with locking jaws in the centre of the fifth wheel [18]. The trailer kingpin is mounted in the trailer upper coupler assembly. The upper coupler consists of the kingpin and the bolster plate.

When the vehicle makes different manoeuvres (starting to go uphill or downhill, and during cornering) [18], the fifth wheel allows the free movement of the trailer and more flexibility of the chassis, as shown in **Figures 12–14**.

Rotation about the longitudinal axis of up to 3° of movement between the tractor and trailer is permitted. On a standard fifth wheel, this occurs as a result of clearance between the fifth wheel to bracket fit, compression of the rubber bushes, and also the vertical movement between the kingpin and locks may allow some lift of the trailer to one side [18].

Consider the third movement of the trailer, the mechanism that represents the fifth wheel has similar design and movements to the suspension mechanism (**Figure 15**), it is located over the front suspension mechanism.

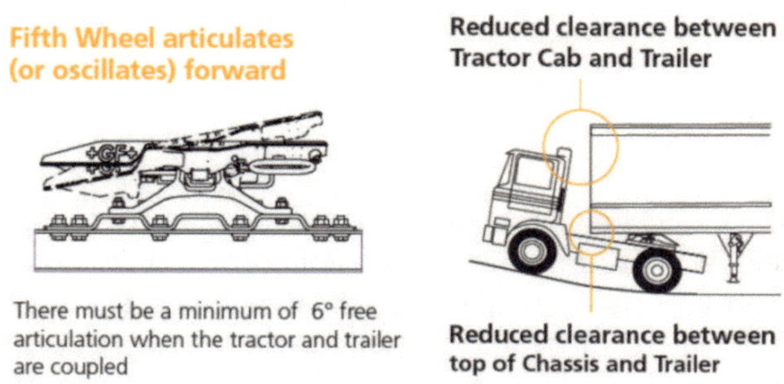

Figure 12.
Movement of the fifth wheel—Starting uphill. Source: Adapted from Saf-Holland [18].

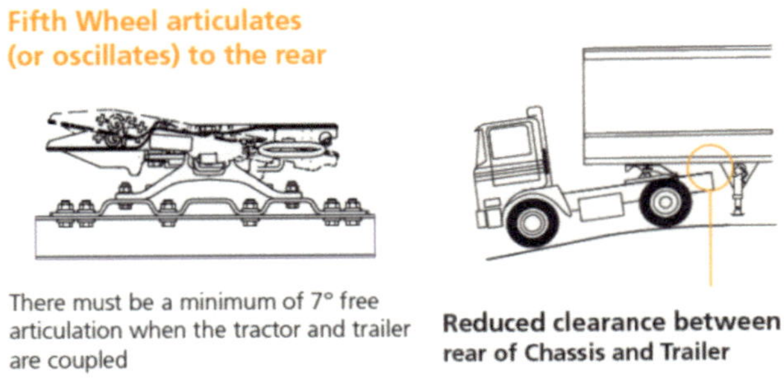

Figure 13.
Movement of the fifth wheel—Starting downhill. Source: Adapted from Saf-Holland [18].

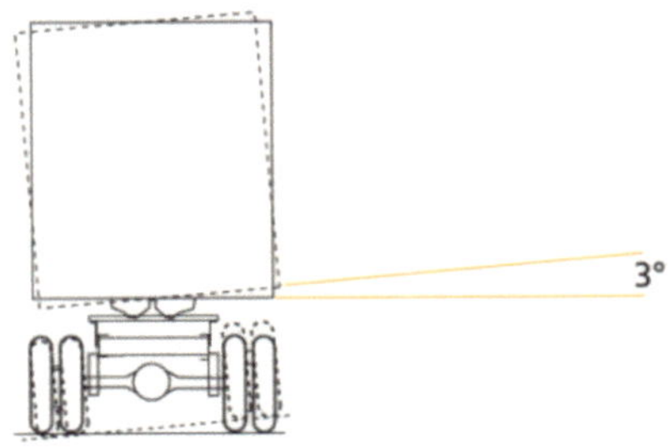

Figure 14.
Movement of the fifth wheel—Rotation x-axis. Source: Adapted from Saf-Holland [18].

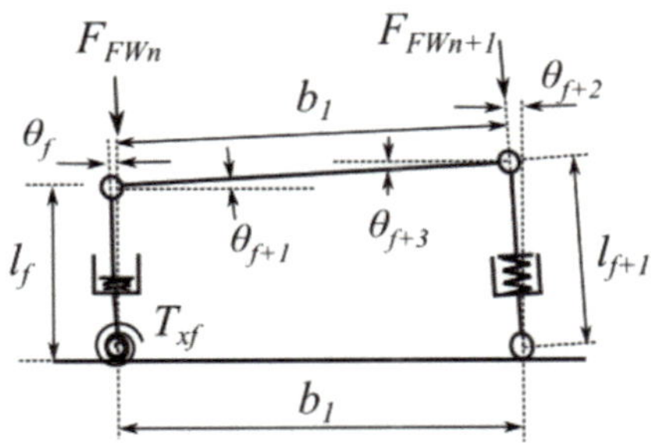

Figure 15.
Kinematic chain of fifth wheel system.

Here l_{fw} is the fifth wheel system's instantaneous height, F_{FWn} is the fifth wheel normal load, and l_{fi} is the fifth wheel system's initial height, b_1 is the fifth wheel width.

1.4 The chassis

The chassis is the backbone of the trailer, and it integrates the main truck component systems such as the axles, suspension, power train, and cab. The chassis is also an important part that contributes to the dynamic performance of the whole vehicle. One of the truck's important dynamic properties is the torsional stiffness, which causes different lateral load transfers (LLT) on the axles of the vehicle [19].

According to Winkler [20] and Rill [4], the chassis has significant torsional compliance, which would allow its front and rear parts to roll independently; this is

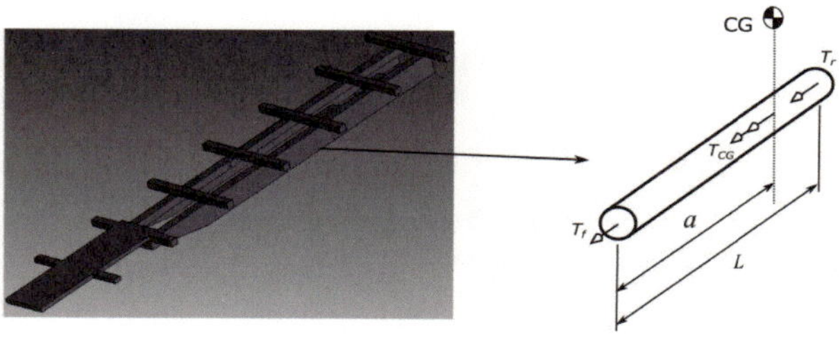

Figure 16.
Kinematic chain of the chassis.

because the lateral load transfer is different on the axles of the vehicle. Then, applying the torsion theory, the vehicle frame has similar behaviour with a statically indeterminate torsional shaft, as shown in **Figure 16**.

Here T_{CG} is the torque applied by the forces acting on the CG, T_f (T_{28}) is the torque applied on the vehicle front axle, T_r (T_{27}) is the torque applied on the vehicle rear axle, a is the distance from the front axle to the centre of gravity, and L is the wheelbase of the trailer. Applying torsion theory to the statically indeterminate shaft, the next equation is defined:

$$\frac{T_f a}{J_f G} = \frac{T_r (L-a)}{J_r G} \tag{13}$$

where J_f and J_r are the equivalent polar moments on front and rear sections of the vehicle frame respectively, and G is the modulus of rigidity (or shear modulus).

According to Kamnik et al. [21] when a trailer model makes a spiral manoeuvre, the LLT on the rear axle is greater than the LLT on the front axle; therefore the equivalent polar moment on the rear (J_r) is greater than the equivalent polar moment on the front (J_f). These can be expressed as $J_r = x J_f$ (where x is the constant that allows controlling the torque distribution of the chassis); replacing and simplifying Eq. (13):

$$T_f + T_r\left(\frac{a-L}{ax}\right) = 0 \tag{14}$$

However, when the trailer model makes a turn, the torque applied on the vehicle front axle has two components, as shown in **Figure 17** and Eqs. (15) and (16).

$$T_{fx} = T_f \cos\psi \tag{15}$$

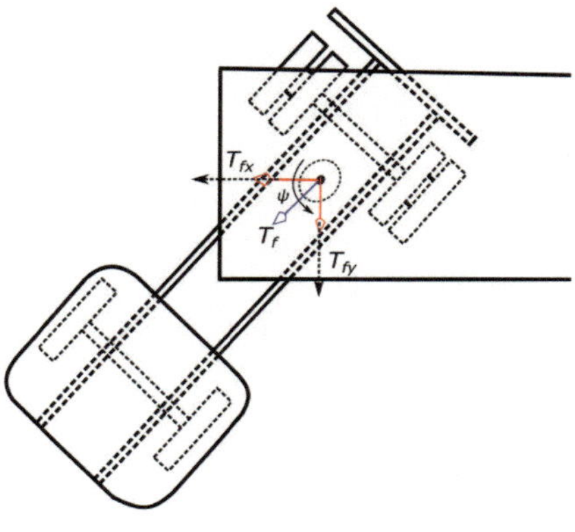

Figure 17.
Torque components.

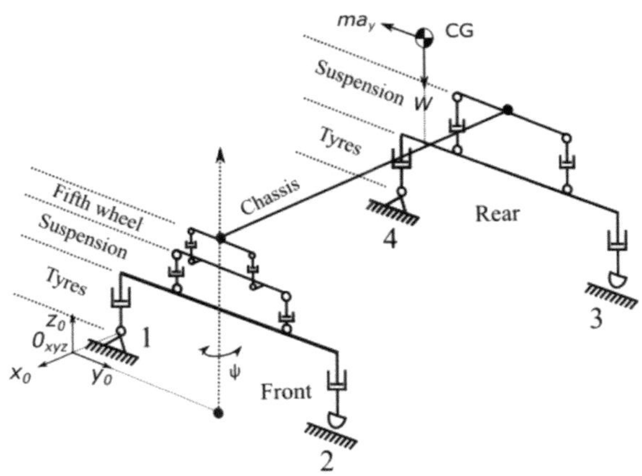

Figure 18.
Trailer model.

$$T_{fy} = T_f \cos \psi \qquad (16)$$

where T_{fx} (T_{x28}) is the torque applied in the x-axis (this torque acts on the lateral load transfer on front axle), T_{fy} (T_{y28}) is the torque applied in the y-axis, and ψ is the steering angle of trailer front axle.

1.5 Three-dimensional trailer model

Considering the systems developed, the model of the trailer (**Figure 18**) is composed of the following mechanisms:

- the front mechanism of the trailer is composed for the tyres, the suspension, and the fifth wheel,
- the rear mechanism is composed for the tyres, and the suspension, and
- the last mechanism is the chassis and links the front and rear mechanism of the model.

The kinematic chain of the trailer model (**Figure 18**) is composed of 28 joints (j = 28; 14 revolute joints 'R', 10 prismatic joints 'P', 2 spherical joints 'S', 2 spherical slider joints 'S_d'), and 23 links (n = 23).

2. Static analysis of the mechanism

Several methodologies allow us to obtain a complete static analysis of the mechanism. For this purpose, the Davies method was used to analyse the mechanisms statically [11, 22–29]. This method was selected because it offers a straightforward way to obtain a static model of the mechanism, and this model can be easily adaptable using this approach.

2.1 External forces and load distribution

In the majority of *LCVs*, the load on the trailers is fixed and nominally centred; for this reason, the initial position lateral of the centre of gravity is centred and symmetric.

Usually, the national regulation boards establish the maximum load capacity of the axles of *LCVs*; this is based on the design load capacity of the pavement and bridges, so each country has its regulations. In this scope, the designers develop their products considering that the trailer is loaded uniformly, causing the axle's load distribution to be in accordance with the laws. **Figure 19** shows the example of the normal load distribution.

However, some loading does not properly distribute the load, which ultimately changes the centre of gravity of the trailer forwards or backwards, as shown in **Figure 20** respectively.

In **Figures 19** and **20**, F_f and F_r are the forces acting on the front and rear axles respectively.

Generally, the *CG* position is dependent on the type of cargo, and the load distribution on the trailer and it varies in three directions: longitudinal (x-axis), lateral (y-axis), and vertical (z-axis), as shown **Figure 21**.

Here, d_1 denotes the lateral *CG* displacement, d_2 denotes the longitudinal *CG* displacement, and d_3 the vertical *CG* displacement.

Furthermore, **Figure 22a** and **b** show that only the weight (W) and the lateral inertial force (ma_y) act on the trailer *CG*, but, when the model takes into account

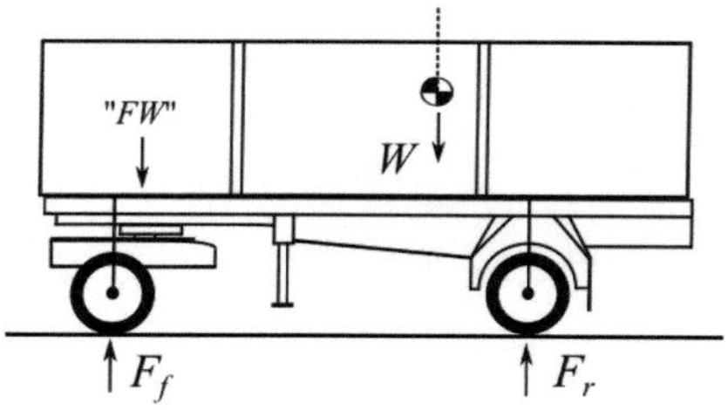

Figure 19.
Normal load distribution.

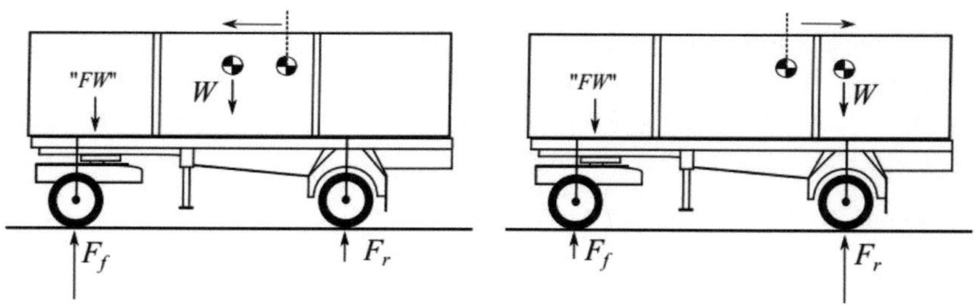

Figure 20.
Longitudinal CG *movement.*

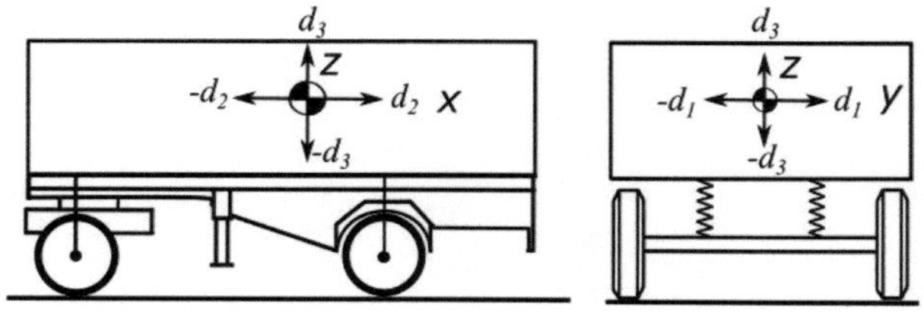

Figure 21.
CG *displacements.*

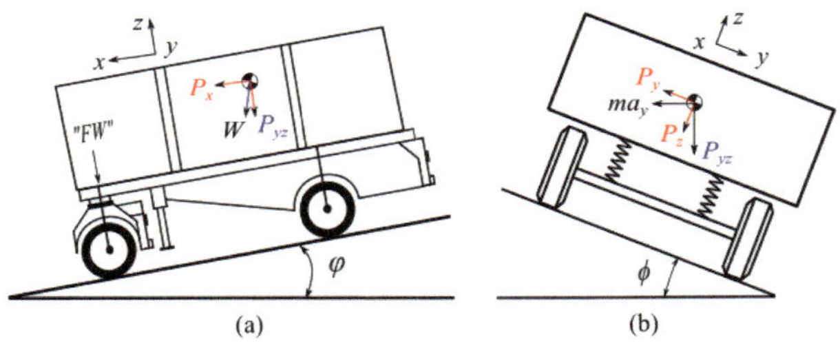

Figure 22.
(a) Longitudinal slope of the road. (b) Banked road.

the longitudinal slope angle (φ) and the bank angle (ϕ) of the road, these forces have three components, as represented in Eqs. (17)–(19).

$$P_x = W \sin \varphi \tag{17}$$

$$P_y = -W \sin \phi \cos \varphi + m a_y \cos \phi \tag{18}$$

$$P_z = W \cos \phi \cos \varphi + m a_y \sin \phi \tag{19}$$

where P_x is the force acting on the x-axis, P_y is the force acting on the y-axis, and P_z is the force acting on the z-axis.

Finally, the load distribution of the trailer on a road with a slope angle is given by the **Figure 23** and Eq. (20).

$$P_x h_2 - P_z(a \pm d_2) + F_r L = 0 \tag{20}$$

where h_2 is the instantaneous CG height, L is the wheelbase of the trailer, and a is the distance from the front axle to the centre of gravity.

2.2 Screw theory of the mechanism

Screw theory enables the representation of the mechanism's instantaneous position in a coordinate system (successive screw displacement method) and the representation of the forces and moments (wrench), replacing the traditional vector representation. All these fundamentals applied to the mechanism are briefly presented below.

2.2.1 Method of successive screw displacements of the mechanism

In the kinematic model for a mechanism, the successive screws displacement method is used. **Figures 24–28** and **Table 1** present the screw parameters of the mechanism.

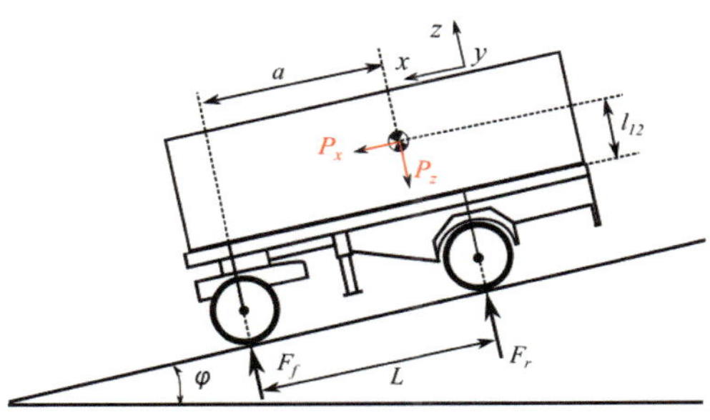

Figure 23.
Load distribution of a trailer on a road with slope angle.

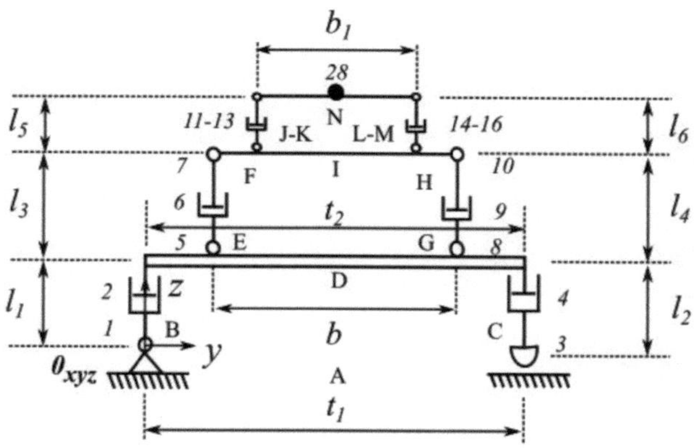

Figure 24.
Variables of the mechanism position (model of the front of the trailer).

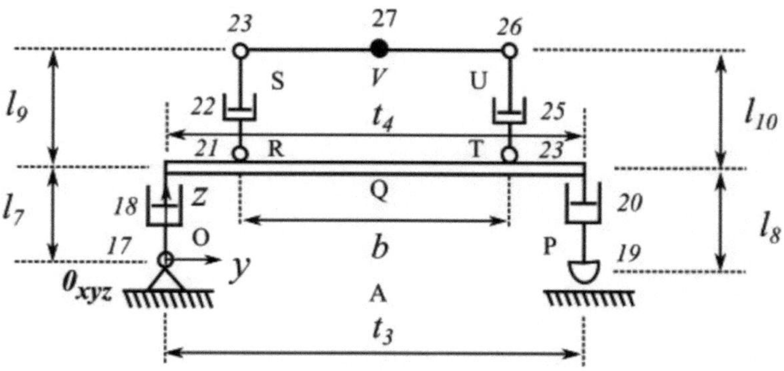

Figure 25.
Variables of the mechanism position (model of the rear of the trailer).

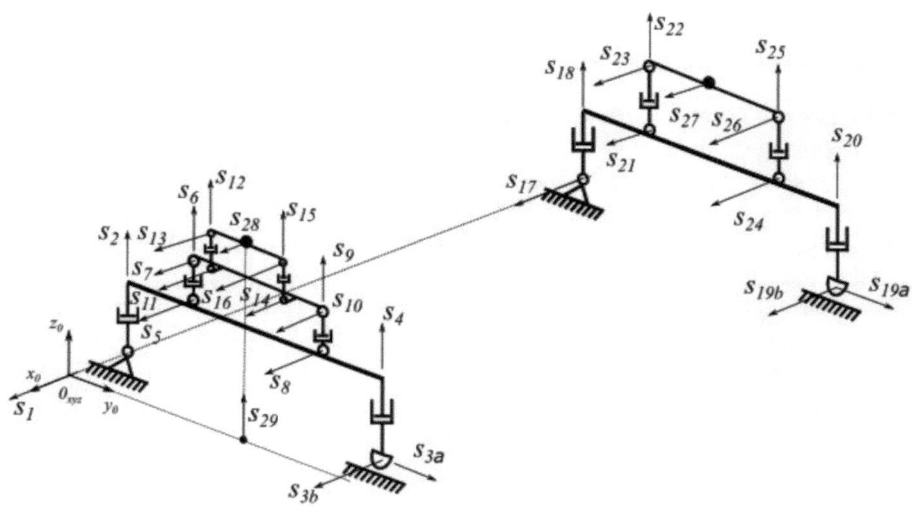

Figure 26.
Vector along the direction of the screws axis (model of the front and rear of the trailer).

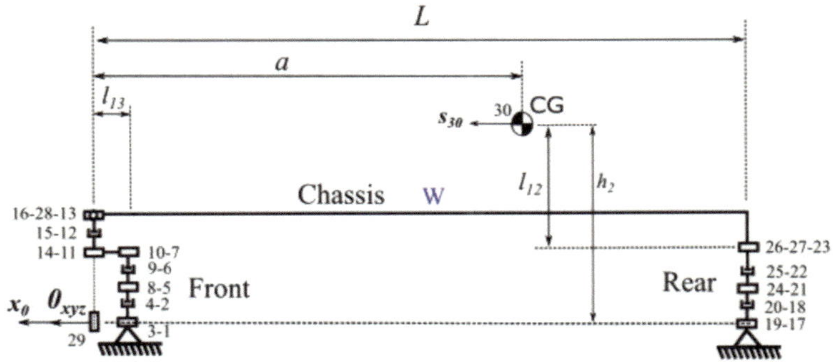

Figure 27.
Variables of the mechanism position (side view of the trailer).

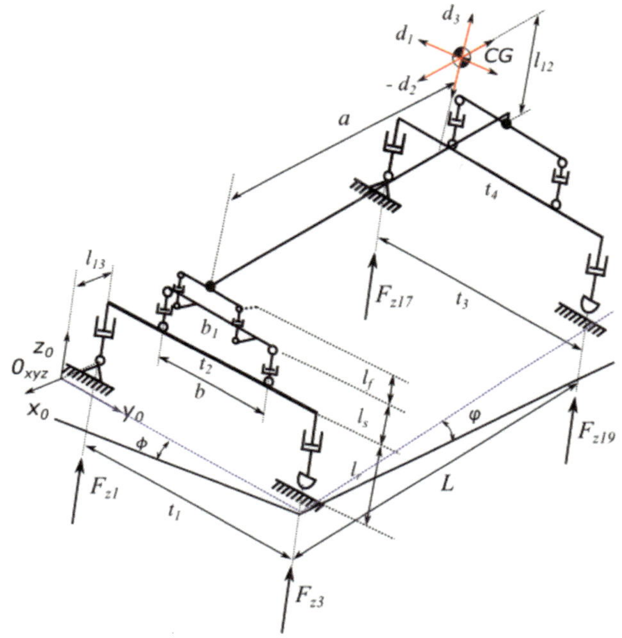

Figure 28.
Variables of the mechanism position (three-dimensional model).

In **Figures 24–28** and **Table 1**, l_{13} is the distance between the fifth wheel and the front axle, $l_{1;2;7;8}$ are the dynamic rolling radii of tyres, $t_{1;3}$ are the front and rear track widths of the trailer respectively, $t_{2;4}$ are the front and rear axle widths respectively, b is the lateral separation between the springs, b_1 is the fifth wheel width, θ_i is the revolution joint angle rotation i, $l_{3;4;9;10}$ are the instantaneous heights of the leaf spring, l_{12} is the height of CG above the chassis, and ψ is the trailer/trailer angle.

This method enables the determination of the displacement of the mechanism and the instantaneous position vector s_{0i} of the joints, and the centre of gravity (The vector s_{0i} (**Table 2**) is obtained from the first three terms of the last column of equations shown in **Table 3**).

2.2.2 Wrench—Forces and moments

In the static analysis, all forces and moments of the mechanism are represented by wrenches ($\$^A$) [13]. The wrenches applied can be represented by the

Joints and points		s			s_O		θ	d
Joint 1	1	0	0	$-l_{13}$	0	0	θ_1	0
Joint 2	0	0	1	$-l_{13}$	0	0	0	l_1
Joint 3a	0	1	0	$-l_{13}$	0	0	0	t_1
Joint 3b	1	0	0	$-l_{13}$	0	0	θ_3	0
Joint 4	0	0	1	$-l_{13}$	0	0	0	l_2
Joint 5	1	0	0	$-l_{13}$	$(t_2 - b)/2$	0	θ_5	0
Joint 6	0	0	1	$-l_{13}$	$(t_2 - b)/2$	0	0	l_3
Joint 7	1	0	0	$-l_{13}$	$(t_2 - b)/2$	0	θ_7	0
Joint 8	1	0	0	$-l_{13}$	$(t_2 + b)/2$	0	θ_8	0
Joint 9	0	0	1	$-l_{13}$	$(t_2 + b)/2$	0	0	l_4
Joint 10	1	0	0	$-l_{13}$	$(t_2 + b)/2$	0	θ_{10}	0
Joint 11	1	0	0	0	$(t_2 - b_1)/2$	0	θ_{11}	0
Joint 12	0	0	1	0	$(t_2 - b_1)/2$	0	0	l_5
Joint 13	1	0	0	0	$(t_2 - b_1)/2$	0	θ_{13}	0
Joint 14	1	0	0	0	$(t_2 + b_1)/2$	0	θ_{14}	0
Joint 15	0	0	1	0	$(t_2 + b_1)/2$	0	0	l_6
Joint 16	1	0	0	0	$(t_2 + b_1)/2$	0	θ_{16}	0
Joint 17	1	0	0	$-L$	0	0	θ_{17}	0
Joint 18	0	0	1	$-L$	0	0	0	l_7
Joint 19a	0	1	0	$-L$	0	0	0	t_3
Joint 19b	1	0	0	$-L$	0	0	θ_{19}	0
Joint 20	0	0	1	$-L$	0	0	0	l_8
Joint 21	1	0	0	$-L$	$(t_4 - b)/2$	0	θ_{21}	0
Joint 22	0	0	1	$-L$	$(t_4 - b)/2$	0	0	l_9
Joint 23	1	0	0	$-L$	$(t_4 - b)/2$	0	θ_{23}	0
Joint 24	1	0	0	$-L$	$(t_4 + b)/2$	0	θ_{24}	0
Joint 25	0	0	1	$-L$	$(t_4 + b)/2$	0	0	l_{10}
Joint 26	1	0	0	$-L$	$(t_4 + b)/2$	0	θ_{26}	0
Joint 27	1	0	0	$-L$	$t_4/2$	0	θ_{27}	0
Joint 28	1	0	0	0	$t_2/2$	0	θ_{28}	0
Point 29	0	0	1	0	$t_2/2$	0	ψ	0
CG (30)	1	0	0	$-a \pm d_2$	$(t_4/2) \pm d_1$	$l_{12} \pm d_3$	0	0

Table 1.
Screw parameters of the mechanism.

vector $\$^A = [M_x\, M_y\, M_z\, F_x\, F_y\, F_z]^T$, where F denotes the forces, and M denotes the moments.

To simplify the model of **Figure 28**, the following considerations were made:

- for the x-direction a steady-state model was used in the analysis;
- disturbances imposed by the road and the lateral friction forces (F_y) (tyre-ground contact) in the joints 3 and 19 were neglected; and

References	s_{Oi}		
1	$\frac{t_2 s_{29} - 2l_{13} c_{29}}{2}$	$-\frac{2l_{13} s_{29} + t_2 c_{29} - t_2}{2}$	0
2	$\frac{(2l_1 s_1 + t_2) s_{29} - 2l_{13} c_{29}}{2}$	$-\frac{(2l_1 s_1 + t_2) c_{29} + 2l_{13} s_{29} - t_2}{2}$	$l_1 c_1$
3	$\frac{(t_2 - 2t_1) s_{29} - 2l_{13} c_{29}}{2}$	$-\frac{2l_{13} s_{29} + (t_2 - 2t_1) c_{29} - t_2}{2}$	0
4	$\frac{(2l_2 s_3 + t_2 - 2t_1) s_{29} - 2l_{13} c_{29}}{2}$	$-\frac{2l_{13} s_{29} + (2l_2 s_3 + t_2 - 2t_1) c_{29} - t_2}{2}$	$l_2 c_3$
5–16	⋮	⋮	⋮
17	$-L$	0	0
18	$-L$	$-l_7 s_{17}$	$l_7 c_{17}$
19	$-L$	t_3	0
20	$-L$	$t_3 - l_8 s_{19}$	$l_8 c_{19}$
21	$-L$	$-\frac{2l_{17} s_{17} + (b_2 - t_4) c_{17}}{2}$	$\frac{(t_4 - b_2) s_{17} + 2l_{17} c_{17}}{2}$
22–23	$-L$	$-\frac{2l_{17} s_{17} + (b_2 - t_4) c_{17} + 2l_9 s_{21+17}}{2}$	$\frac{(t_4 - b_2) s_{17} + 2l_{17} c_{17} + 2l_9 c_{21+17}}{2}$
24–28	⋮	⋮	⋮
CG	$-a \pm d_2$	$^*h_1$	$^*h_2$

$s_i = \sin\theta_i c_i = \cos\theta_i$

$^*h_1 = -\frac{(2l_{12} \pm 2d_3) s_{27+23+21+17} \pm d_1 c_{27+23+21+17} - b_2 c_{23+21+17} + 2l_9 s_{21+17} + 2l_7 s_{17} + (b_2 - t_4) c_{17}}{2.}$

$^*h_2 = -\frac{\pm d_1 s_{27+23+21+17} + (\pm 2d_3 - 2l_{12}) c_{27+23+21+17} - b_2 s_{23+21+17} - 2l_9 c_{21+17} - 2l_7 c_{17} + (b_2 - t_4) s_{17}}{2}$

Table 2.
Instantaneous position vector s_{Oi}.

Joints and points	Instantaneous position matrix
Joint 1	$p'_1 = A_{29} A_1 p_1$
Joint 2	$p'_2 = A_{29} A_1 A_2 p_2$
Joint 3	$p'_3 = A_{29} A_{3a} A_{3b} p_3$
Joint 4	$p'_4 = A_{29} A_{3a} A_{3b} A_4 p_4$
Joint 5	$p'_5 = A_{29} A_1 A_2 A_5 p_5$
Joint 6	$p'_6 = A_{29} A_1 A_2 A_5 A_6 p_6$
Joint 7	$p'_7 = A_{29} A_1 A_2 A_5 A_6 A_7 p_7$
Joint 8	$p'_8 = A_{29} A_1 A_2 A_8 p_8$
Joint 9	$p'_9 = A_{29} A_1 A_2 A_8 A_9 p_9$
Joint 10	$p'_{10} = A_{29} A_1 A_2 A_8 A_9 A_{10} p_{10}$
Joint 11	$p'_{11} = A_{29} A_1 A_2 A_5 A_6 A_7 A_{11} p_{11}$
Joint 12	$p'_{12} = A_{29} A_1 A_2 A_5 A_6 A_7 A_{11} A_{12} p_{12}$
Joint 13	$p'_{13} = A_{29} A_1 A_2 A_5 A_6 A_7 A_{11} A_{12} A_{13} p_{13}$
Joint 14	$p'_{14} = A_{29} A_1 A_2 A_5 A_6 A_7 A_{14} p_{14}$
Joint 15	$p'_{15} = A_{29} A_1 A_2 A_5 A_6 A_7 A_{14} A_{15} p_{15}$
Joint 16	$p'_{16} = A_{29} A_1 A_2 A_5 A_6 A_7 A_{14} A_{15} A_{16} p_{16}$
Joint 17	$p'_{17} = A_{17} p_{17}$
Joint 18	$p'_{18} = A_{17} A_{18} p_{18}$
Joint 19	$p'_{19} = A_{19a} A_{19b} p_{19}$

Joints and points	Instantaneous position matrix
Joint 20	$p'_{20} = A_{19a}\, A_{19b}\, A_{20}\, p_{20}$
Joint 21	$p'_{21} = A_{17}\, A_{18}\, A_{21}\, p_{21}$
Joint 22	$p'_{22} = A_{17}\, A_{18}\, A_{21}\, A_{22}\, p_{22}$
Joint 23	$p'_{23} = A_{17}\, A_{18}\, A_{21}\, A_{22}\, A_{23}\, p_{23}$
Joint 24	$p'_{24} = A_{17}\, A_{18}\, A_{24}\, p_{24}$
Joint 25	$p'_{25} = A_{17}\, A_{18}\, A_{24}\, A_{25}\, p_{25}$
Joint 26	$p'_{26} = A_{17}\, A_{18}\, A_{24}\, A_{25}\, A_{26}\, p_{26}$
Joint 27	$p'_{27} = A_{17}\, A_{18}\, A_{21}\, A_{22}\, A_{23}\, A_{27}\, p_{27}$
Joint 28	$p'_{28} = A_{29}\, A_{1}\, A_{2}\, A_{5}\, A_{6}\, A_{7}\, A_{11}\, A_{12}\, A_{13}\, A_{28}\, p_{28}$
CG (30)	$p'_{CG} = A_{17}\, A_{18}\, A_{21}\, A_{22}\, A_{23}\, A_{27}\, A_{CG}\, p_{CG}$

Table 3.
Instantaneous position matrix.

Joints and reference points	Constraints and forces		s_i		Inst. position vector s_{0i}
Revolute joints 1, 7, 8, 10, 13, 14, 16, 17, 23, 24, and 26	F_{xi}	1	0	0	Revolute joints 1, 7, 8, 10, 13, 14, 16, 17, 23, 24, and 26
	F_{yi}	0	1	0	
	F_{zi}	0	0	1	
	M_{yi}	0	1	0	
	M_{zi}	0	0	1	
Spherical slider joints 3 and 19	F_{xi}	1	0	0	Spherical slider joints 3 and 19
	F_{zi}	0	0	1	
Revolute joints 5, 11, and 21	F_{xi}	1	0	0	Revolute joints 5, 11, and 21
	F_{yi}	0	1	0	
	F_{zi}	0	0	1	
	T_{xi}	1	0	0	
	M_{yi}	0	1	0	
	M_{zi}	0	0	1	
Prismatic joints 2, 4, 6, 9, 12, 15, 18, 20, 22 and 25	F_{xi}	1	0	0	Prismatic joints 2, 4, 6, 9, 12, 15, 18, 20, 22, and 25
	F_{ni}	0	$\cos\theta_{i-1}$	$\sin\theta_{i-1}$	
	M_{xi}	1	0	0	
	M_{yi}	0	1	0	
	M_{zi}	0	0	1	
Prismatic joints 2, 4, 18, and 20	F_{Ti}	0	$-\sin\theta_{i-1}$	$\cos\theta_{i-1}$	Prismatic joints 2, 4, 18, and 20
Prismatic joints 6, 9, 22, and 25	F_{LSi}	0	$-\sin\theta_{i-1}$	$\cos\theta_{i-1}$	Prismatic joints 6, 9, 22, and 25
Prismatic joints 12 and 15	F_{FWi}	0	$-\sin\theta_{i-1}$	$\cos\theta_{i-1}$	Prismatic joints 12 and 15

Joints and reference points	Constraints and forces	s_i			Inst. position vector s_{0i}
Spherical joints 27 and 28	F_{xi}	1	0	0	Spherical joints 27 and 28
	F_{yi}	0	1	0	
	F_{zi}	0	0	1	
	T_{xi}	1	0	0	
CG (30)	P_x	1	0	0	CG (30)
	P_y	0	−1	0	
	P_z	0	0	−1	

Table 4.
Wrench parameters of the mechanism.

- the components of the trailer weight (W) and the inertial force (ma_y) are the only external forces acting on the trailer CG.

Considering a static analysis in a three-dimensional space [7], the corresponding wrenches of each joint and external forces are defined by the parameters of **Table 4**, where s_i represents the orientation vector of each wrench i.

All of the wrenches of the mechanism together comprise the action matrix $[A_d]$ given by Eq. (21) (or the amplified matrix of the Eq. (22)).

$$[A_d]_{6\times148} = \left[\$^A_{F_{x1}}\ \$^A_{F_{y1}}\ \$^A_{F_{z1}}\ \cdots\ \$^A_{P_x}\ \$^A_{P_y}\ \$^A_{P_z}\right] \tag{21}$$

$$[A_d] = \begin{bmatrix} 0 & 0 & -p_1F_{z1} & \cdots & 0 & h_2P_y & -h_1P_z \\ 0 & 0 & -p_2F_{z1} & \cdots & h_2P_x & 0 & (-a \pm d_2)P_z \\ p_1F_{x1} & p_2F_{y1} & 0 & \cdots & -h_1P_x & -(-a \pm d_2)P_y & 0 \\ F_{x1} & 0 & 0 & \cdots & P_x & 0 & 0 \\ 0 & F_{y1} & 0 & \cdots & 0 & -P_y & 0 \\ 0 & 0 & F_{z1} & \cdots & 0 & 0 & -P_z \end{bmatrix} \tag{22}$$

where p_i is a system variable.

The wrench can be represented by a normalised wrench and a magnitude. Therefore, from the Eq. (22) the unit action matrix and the magnitudes action vector are obtained, as represented by Eqs. (23) and (24).

$$\left[\hat{A}_d\right]_{6\times148} = \begin{bmatrix} 0 & 0 & -p_1 & \cdots & 0 & h_2 & -h_1 \\ 0 & 0 & -p_2 & \cdots & h_2 & 0 & (-a \pm d_2) \\ p_1 & p_2 & 0 & \cdots & -h_1 & -(-a \pm d_2) & 0 \\ 1 & 0 & 0 & \cdots & 1 & 0 & 0 \\ 0 & 1 & 0 & \cdots & 0 & -1 & 0 \\ 0 & 0 & 1 & \cdots & 0 & 0 & -1 \end{bmatrix} \tag{23}$$

$$[\Psi]_{148\times1} = \left[F_{x1}\ \ F_{y1}\ \ F_{z1}\ \ \cdots\ \ P_x\ \ P_y\ \ P_z\right] \tag{24}$$

2.2.3 Graph theory

Kinematic chains and mechanisms are comprised of links and joints, which can be represented by graphs, where the vertices correspond to the links, and the edges correspond to the joints and external forces [5, 7].

The mechanism of the **Figure 28** is represented by the direct coupling graph of the **Figure 29**. This graph has 23 vertices (links) and 31 edges (joints and external forces (P_x, P_y, and P_z)).

The direct coupling graph (**Figure 29**) can be represented by the incidence matrix $[I]_{23X31}$ [30] (Eq. (25)). The incidence matrix provides the mechanism cut-set matrix $[Q]_{22X31}$ [11, 25–28, 30] (Eq. (26)) for the mechanism, where each line represents a cut graph and the columns represent the mechanism joints. Besides, this matrix is rearranged, allowing 22 branches (edges *1–3, 5–9, 11–15, 17–19, 21–25, and 27*—identity matrix) and 9 chords (edges *4, 10, 16, 20, 26, 28,* P_x, P_y, and P_z) to be defined, as shown in **Figure 30**.

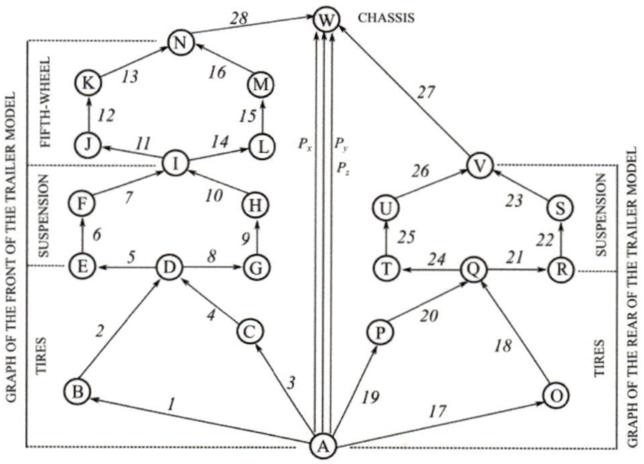

Figure 29.
Direct coupling graph of the mechanism.

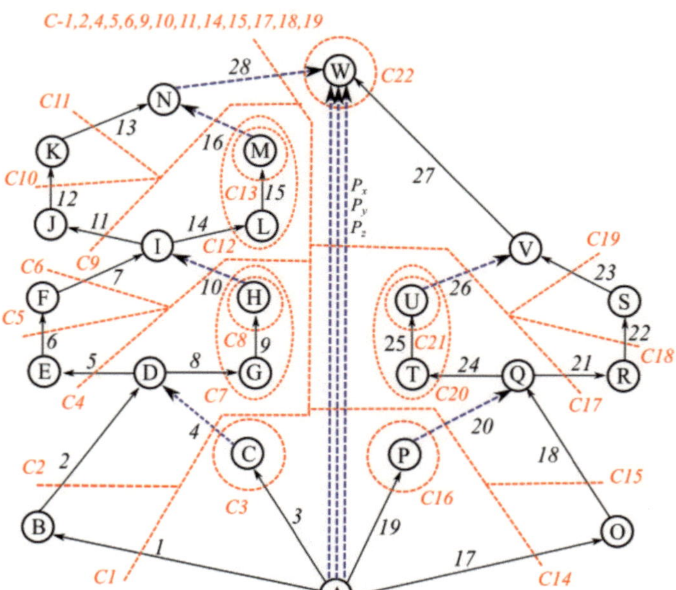

Figure 30.
Cut-set action graph of the mechanism.

$[I]_{23X31} =$

	1	2	3	4	5	6	7	8	...	...	25	26	27	28	P_x	P_z	P_y
A	1	0	1	0	0	0	0	0	...	...	0	0	0	0	1	1	1
B	-1	1	0	0	0	0	0	0	...	...	0	0	0	0	0	0	0
C	0	0	-1	1	0	0	0	0	...	...	0	0	0	0	0	0	0
D	0	-1	0	-1	1	0	0	1	...	...	0	0	0	0	0	0	0
E	0	0	0	0	-1	1	0	0	...	...	0	0	0	0	0	0	0
F	0	0	0	0	0	-1	1	0	...	...	0	0	0	0	0	0	0
G	0	0	0	0	0	0	0	-1	...	...	0	0	0	0	0	0	0
H	0	0	0	0	0	0	0	0	...	...	0	0	0	0	0	0	0
I	0	0	0	0	0	0	-1	0	...	...	0	0	0	0	0	0	0
J	0	0	0	0	0	0	0	0	...	...	0	0	0	0	0	0	0
K	0	0	0	0	0	0	0	0	...	...	0	0	0	0	0	0	0
L	0	0	0	0	0	0	0	0	...	...	0	0	0	0	0	0	0
M	0	0	0	0	0	0	0	0	...	...	0	0	0	0	0	0	0
N	0	0	0	0	0	0	0	0	...	...	0	0	0	1	0	0	0
O	0	0	0	0	0	0	0	0	...	...	0	0	0	0	0	0	0
P	0	0	0	0	0	0	0	0	...	...	0	0	0	0	0	0	0
Q	0	0	0	0	0	0	0	0	...	...	0	0	0	0	0	0	0
R	0	0	0	0	0	0	0	0	...	...	0	0	0	0	0	0	0
S	0	0	0	0	0	0	0	0	...	...	0	0	0	0	0	0	0
T	0	0	0	0	0	0	0	0	...	...	1	0	0	0	0	0	0
U	0	0	0	0	0	0	0	0	...	...	-1	1	0	0	0	0	0
V	0	0	0	0	0	0	0	0	...	...	0	-1	1	0	0	0	0
W	0	0	0	0	0	0	0	0	...	...	0	0	-1	-1	-1	-1	-1

(25)

$[Q]_{22X31} =$ (CUTS)

	1	2	3	4	5	6	7	8	...	...	25	26	27	28	P_x	P_z	P_y
1	1	0	0	1	0	0	0	0	...	...	0	0	0	-1	0	0	0
2	0	1	0	1	0	0	0	0	...	...	0	0	0	-1	0	0	0
3	0	0	1	-1	0	0	0	0	...	...	0	0	0	0	0	0	0
4	0	0	0	0	1	0	0	0	...	...	0	0	0	-1	0	0	0
5	0	0	0	0	0	1	0	0	...	...	0	0	0	-1	0	0	0
6	0	0	0	0	0	0	1	0	...	...	0	0	0	-1	0	0	0
7	0	0	0	0	0	0	0	1	...	...	0	0	0	0	0	0	0
8	0	0	0	0	0	0	0	0	...	...	0	0	0	0	0	0	0
9	0	0	0	0	0	0	0	0	...	...	0	0	0	-1	0	0	0
10	0	0	0	0	0	0	0	0	...	...	0	0	0	-1	0	0	0
11	0	0	0	0	0	0	0	0	...	...	0	0	0	-1	0	0	0
12	0	0	0	0	0	0	0	0	...	...	0	0	0	0	0	0	0
13	0	0	0	0	0	0	0	0	...	...	0	0	0	0	0	0	0
14	0	0	0	0	0	0	0	0	...	...	0	0	0	1	1	1	1
15	0	0	0	0	0	0	0	0	...	...	0	0	0	1	1	1	1
16	0	0	0	0	0	0	0	0	...	...	0	0	0	0	0	0	0
17	0	0	0	0	0	0	0	0	...	...	0	1	0	1	1	1	1
18	0	0	0	0	0	0	0	0	...	...	0	1	0	1	1	1	1
19	0	0	0	0	0	0	0	0	...	...	0	1	0	1	1	1	1
20	0	0	0	0	0	0	0	0	...	...	0	-1	0	0	0	0	0
21	0	0	0	0	0	0	0	0	...	...	1	-1	0	0	0	0	0
22	0	0	0	0	0	0	0	0	...	...	0	0	1	1	1	1	1

(26)

All constraints are represented as edges, which allows the amplification of the cut-set graph and the cut-set matrix. Additionally, the tyre normal load (F_{Ti}), spring normal load (F_{LSi}), fifth wheel normal load (F_{FWi}), and the passive torsional moment (T_{xi}) are included.

Figure 30 presents the cut-set action graph and the Eq. (27) presents the expanded cut-set matrix ($[Q]_{22X148}$), where each line represents a cut of the graph, and the columns are the constraints of the joints as well as external forces on the mechanism.

$[Q]_{22X148} =$ (CUTS)

F_{x1}	F_{y1}	F_{z1}	M_{y1}	M_{z1}	F_{x2}	F_{p2}	F_{T2}	...	F_{x28}	F_{y28}	F_{z28}	T_{x28}	P_x	P_z	P_y
1	1	1	1	1	0	0	0	...	0	-1	-1	-1	0	0	0
0	0	0	0	0	1	1	1	...	0	-1	-1	-1	0	0	0
0	0	0	0	0	0	0	0	...	0	0	0	0	0	0	0
0	0	0	0	0	0	0	0	...	0	-1	-1	-1	0	0	0
0	0	0	0	0	0	0	0	...	0	-1	-1	-1	0	0	0
0	0	0	0	0	0	0	0	...	0	-1	-1	-1	0	0	0
0	0	0	0	0	0	0	0	...	0	0	0	0	0	0	0
0	0	0	0	0	0	0	0	...	0	0	0	0	0	0	0
0	0	0	0	0	0	0	0	...	0	-1	-1	-1	0	0	0
0	0	0	0	0	0	0	0	...	0	-1	-1	-1	0	0	0
0	0	0	0	0	0	0	0	...	0	-1	-1	-1	0	0	0
0	0	0	0	0	0	0	0	...	0	0	0	0	0	0	0
0	0	0	0	0	0	0	0	...	0	0	0	0	0	0	0
0	0	0	0	0	0	0	0	...	0	1	1	1	1	1	1
0	0	0	0	0	0	0	0	...	0	1	1	1	1	1	1
0	0	0	0	0	0	0	0	...	0	0	0	0	0	0	0
0	0	0	0	0	0	0	0	...	0	1	1	1	1	1	1
0	0	0	0	0	0	0	0	...	0	1	1	1	1	1	1
0	0	0	0	0	0	0	0	...	0	1	1	1	1	1	1
0	0	0	0	0	0	0	0	...	0	0	0	0	0	0	0
0	0	0	0	0	0	0	0	...	0	0	0	0	0	0	0
0	0	0	0	0	0	0	0	...	1	1	1	1	1	1	1

(27)

2.3 Equation system solution

Using the cut-set law [24], the algebraic sum of the normalised wrenches given in Eqs. (23) and (24), that belong to the same cut $[Q]_{22X148}$ (**Figure 30** and Eq. (27))

must be equal to zero. Thus, the statics of the mechanism can be defined, as exemplified in Eq. (28) (or the amplified matrix of the Eq. (29)):

$$\left[\hat{A}_n\right]_{132\times148}[\Psi]^T_{148\times1} = [0]_{132\times1} \tag{28}$$

$$\begin{matrix} \\ \\ \\ Cut\ 1 \\ \\ \\ \vdots \\ \\ \\ Cut\ 22 \\ \\ \\ \\ \end{matrix}\begin{bmatrix} 0 & 0 & \cdots & 0 & 0 & 0 \\ 0 & 0 & \cdots & 0 & 0 & 0 \\ p_1 & p_2 & \cdots & 0 & 0 & 0 \\ 1 & 0 & \cdots & 0 & 0 & 0 \\ 0 & 1 & \cdots & 0 & 0 & 0 \\ 0 & 0 & \cdots & 0 & 0 & 0 \\ \vdots & \vdots & \vdots & \vdots & \vdots & \vdots \\ 0 & 0 & \cdots & 0 & h_2 & -h_1 \\ 0 & 0 & \cdots & h_2 & 0 & (-a \pm d_2) \\ 0 & 0 & \cdots & -h_1 & -(-a \pm d_2) & 0 \\ 0 & 0 & \cdots & 1 & 0 & 0 \\ 0 & 0 & \cdots & 0 & -1 & 0 \\ 0 & 0 & \cdots & 0 & 0 & -1 \end{bmatrix} \cdot \begin{bmatrix} F_{x1} \\ F_{y1} \\ \vdots \\ P_x \\ P_y \\ P_z \end{bmatrix} = [0]_{132\times1} \tag{29}$$

It is necessary to identify the set of primary variables $[\Psi_p]$ (known variables), among the variables of Ψ. Once identified, the system of the Eq. (28) is rearranged and divided in two sets, as shown by Eq. (30).

$$\left[\hat{A}_{ns}\right]_{132\times145}[\Psi_s]^T_{145\times1} + \left[\hat{A}_{np}\right]_{132\times3}\left[\Psi_p\right]^T_{3\times1} = [0]_{132\times1} \tag{30}$$

where $[\Psi_p]$ is the primary variable vector, $[\Psi_s]$ is the second variable vector (unknown variables), $\left[\hat{A}_{np}\right]$ are the columns corresponding to the primary variables, and $\left[\hat{A}_{ns}\right]$ are the columns corresponding to the secondary variables.

In this case, the primary variable vector is:

$$\left[\Psi_p\right]_{3\times1} = \left[P_x \quad P_y \quad P_z\right]^T \tag{31}$$

and the secondary variable vector is:

$$[\Psi_s]_{145\times1} = \left[F_{x1} \quad F_{y1} \quad M_{y1} \quad M_{z1} \quad \cdots \quad F_{z1} \quad F_{z3} \quad F_{z17} \quad F_{z19}\right]^T \tag{32}$$

Solving the system Eq. (30) using the Gauss-Jordan elimination method, all secondary variables being function of the primary variables, the last row of the solution system provides the next equation:

$$F_{z3} + \frac{P_1 + t_3}{t_1 \cos\psi}F_{z19} + \frac{P_1}{t_1 \cos\psi}F_{z17}$$

$$-\frac{h_1 + P_1}{t_1 \cos\psi}P_z + \frac{h_2}{t_1 \cos\psi}P_y = 0 \tag{33}$$

replacing P_y and P_z:

$$F_{z3} + \frac{P_1 + t_3}{t_1 \cos\psi} F_{z19} + \frac{P_1}{t_1 \cos\psi} F_{z17} - \frac{h_1 + P_1}{t_1 \cos\psi} \left(W\cos\phi\cos\varphi + ma_y \sin\phi \right) + \frac{h_2}{t_1 \cos\psi} \left(ma_y \cos\phi - W\sin\phi\cos\varphi \right) = 0 \quad (34)$$

where P_1 is a system variable ($P_1 = (2l_{13}\sin\psi + t_2(\cos\psi - 1))/2$), h_1 is the instantaneous lateral distance between the zero-reference frame and the centre of gravity, and h_2 is the instantaneous CG height (**Table 2**). Simplifying the equation, and making $\tan(\phi) = e$, where e is the tangent of the bank angle, we have:

$$\frac{a_y}{g} = \frac{h_1 \cos\varphi + h_2 e\cos\varphi}{h_2 - (h_1 + P_1)e} \times \left(1 - \frac{t_1 F_{z3} \cos\psi + P_1(F_{z17} - W\cos\phi\cos\varphi) + (P_1 + t_3)F_{z19}}{W\cos\phi(h_1 \cos\varphi + h_2 e\cos\varphi)} \right) \quad (35)$$

According to the static redundancy problem known as the four-legged table [31, 32], a plane is defined by just three points in space and, consequently, a four-legged table has support plane multiplicities. This is why when one leg loses contact with the ground, the table is supported by the other three, as shown in **Figure 31**.

The problem of the four-legged table is observed in dynamic rollover tests when the rear inner tyre loses contact with the ground ($F_{z19} = 0$), and the front inner tyre (F_{z3}) does not, as shown, for example, in **Figure 32**.

Applying this theory to the vehicle stability, when a vehicle makes a turn, it is subjected to an increasing lateral load until it reaches the rollover threshold [32]. During the turning, the rear inner tyre is usually the one that loses contact with the ground. For this condition ($F_{z19} = 0$), and thus:

$$SRT_{3D_{\psi\phi\varphi}} = \frac{h_1 \cos\varphi + h_2 e\cos\varphi}{h_2 - (h_1 + P_1)e} \times \left(1 - \frac{t_1 F_{z3} \cos\psi + P_1(F_{z17} - W\cos\phi\cos\varphi)}{W\cos\phi(h_1 \cos\varphi + h_2 e\cos\varphi)} \right) \quad (36)$$

where $SRT_{3D_{\psi\phi\varphi}}$ factor is the three-dimensional static rollover threshold for a trailer model with trailer/trailer angle (ψ), bank angle (e), and slope angle (φ).

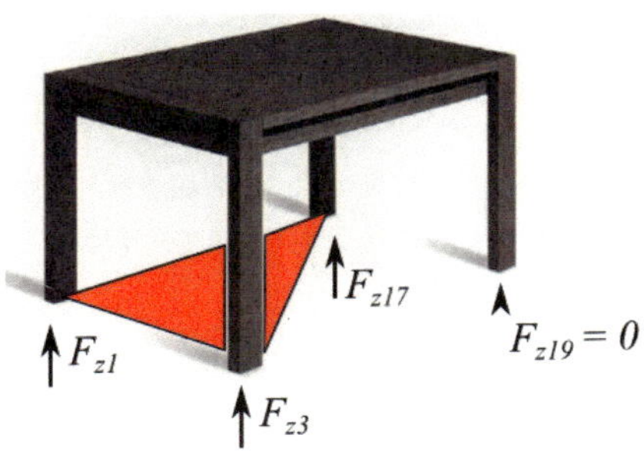

Figure 31.
Redundancy problem of the four-legged table.

Figure 32.
Dynamic rollover test. Source: Adapted of Cabral [33].

The normal forces F_{z3} and F_{z17} depend on the LLT coefficient in the front and rear axles respectively [4, 21, 34]. Furthermore, this coefficient depends on the vehicle type, speed, suspension, tyres, etc.

This information demonstrates that the $SRT_{3D_{\psi\phi\varphi}}$ factor of a vehicle (Eq. (36)) is, in general, inferior to the SRT factor for a two-dimensional model vehicle [35], as shown in Eq. (37).

$$SRT_{2D} = \frac{a_y}{g} = \frac{t}{2h} \tag{37}$$

where h is the CG height, t is the vehicle track.

Parameters of the trailer	**Value**	**Units**
Trailer weight—W	355.22	kN
Front and rear track widths ($t_{1,3}$)	1.86	m
Front and rear axles widths ($t_{2,4}$)	1.86	m
Stiffness of the suspension per axle (k_s) [37]	1800	kN.m^{-1}
Number of axles at the front (trailer) (four tyres per axle)	2	
Number of axles at the rear (trailer) (four tyres per axle)	3	
Vertical stiffness per tyre (k_T) ([37])	840	kN.m^{-1}
Initial suspension height ($l_{3,4,9,10}$) (l_s)	0.205	m
Initial dynamic rolling radius ($l_{1,2,7,8}$) (l_r) (Michelin XZA® [36])	0.499	m
Initial height of the fifth wheel (l_{fi})	0.1	m
Lateral separation between the springs (b)	0.95	m
Fifth wheel width (b_1)	0.6	m
CG height above the chassis (l_{12})	1.346	m
Distance between the fifth wheel and the front axle (l_{13})	0.15	m
Wheelbase of the trailer (L)	4.26	m
Distance from the front axle to the centre of gravity (a)	3	m
Offset of the cargo d_1	0.1	m
Trailer/trailer angle (ψ)	0	°

Table 5.
Parameters of the trailer model.

With Eq. (36), it is possible to obtain a better vehicle stability representation and the $SRT_{3D_{\psi\phi\varphi}}$ factor value attainments closer to reality.

To simplify the solution of the system of equations in Eq. (30), the following hypotheses were considered:

- in the majority of *LCVs*, the load on the trailers is uniformly distributed (Eq. (20));
- the lateral load transfer of the trailer model is controlled through the torsional moment of the chassis (spherical joints 27 and 28 (Eqs. (15) and (16))).

Eq. (38) shows the final static system for the stability analysis, solving this system using the Gauss-Jordan elimination method, all secondary variables are a function of primary variables, (P_x—force acting on the x-axis, P_y—force acting on the y-axis, and P_z—force acting on the z-axis).

$$
\begin{array}{r}
\\ \\ \\ Cut\ 1 \\ \\ \\ \vdots \\ \\ \\ Cut\ 22 \\ \\ \\ Eq.(20) \\ Eq.(15) \\ Eq.(16) \\ \\
\end{array}
\begin{bmatrix}
0 & 0 & \cdots & 0 & 0 & -p_1 & 0 & 0 & 0 & 0 & 0 & 0 \\
0 & 0 & \cdots & 0 & 0 & -p_2 & 0 & 0 & 0 & 0 & 0 & 0 \\
p_1 & p_2 & \cdots & 0 & 0 & 0 & 0 & 0 & 0 & 0 & 0 & 0 \\
1 & 0 & \cdots & 0 & 0 & 0 & 0 & 0 & 0 & 0 & 0 & 0 \\
0 & 1 & \cdots & 0 & 0 & 0 & 0 & 0 & 0 & 0 & 0 & 0 \\
0 & 0 & \cdots & 0 & 0 & 1 & 0 & 0 & 0 & 0 & 0 & 0 \\
\vdots & \vdots & \vdots & \vdots & \vdots & \vdots & \vdots & \vdots & \vdots & \vdots & \vdots & \vdots \\
0 & 0 & \cdots & 0 & 0 & 0 & 0 & 0 & 0 & 0 & h_2 & -h_1 \\
0 & 0 & \cdots & 0 & 0 & 0 & 0 & 0 & 0 & h_2 & 0 & (-a \pm d_2) \\
0 & 0 & \cdots & 0 & 0 & 0 & 0 & 0 & 0 & -h_1 & -(-a \pm d_2) & 0 \\
0 & 0 & \cdots & 0 & 0 & 0 & 0 & 0 & 0 & 1 & 0 & 0 \\
0 & 0 & \cdots & 0 & 0 & 0 & 0 & 0 & 0 & 0 & -1 & 0 \\
0 & 0 & \cdots & 0 & 0 & 0 & 0 & 0 & 0 & 0 & 0 & -1 \\
0 & 0 & \cdots & 0 & 0 & 0 & 0 & L & L & h_2 & 0 & -(-a \pm d_2) \\
0 & 0 & \cdots & 0 & 0 & 0 & 0 & 0 & 0 & 0 & 0 & 0 \\
0 & 0 & \cdots & 0 & 0 & 0 & 0 & 0 & 0 & 0 & 0 & 0
\end{bmatrix}
\cdot
\begin{bmatrix}
F_{x1} \\ F_{y1} \\ \vdots \\ T_{x5} \\ T_{x11} \\ T_{x21} \\ T_{x27} \\ T_{x28} \\ F_{T2} \\ F_{T4} \\ F_{T18} \\ F_{T20} \\ F_{LS6} \\ F_{LS9} \\ F_{LS22} \\ F_{LS25} \\ F_{FW12} \\ F_{FW17} \\ F_{z1} \\ F_{z3} \\ F_{z19} \\ F_{z17} \\ P_x \\ P_y \\ P_z
\end{bmatrix}
= [0]_{135\times 1} \quad (38)
$$

3. Case study

In this study, a B-train trailer with two axles on front and three axles on the rear is analysed. This model has a suspension system with a tandem axle, and its parameters depend on the construction materials. Another important parameter of the model is the dynamic rolling radius or loaded radius l_i. The proposed model considers Michelin XZA® [36] radial tyres. **Table 5** shows the parameters of the trailer used in this analysis [32, 38].

To calculate the *SRT* factor, the inertial force is increased until the lateral load transfer in the rear axle is complete (the entire load is transferred from the rear inner tyre to the rear outer tyre when the model makes a turn). The reduction in the *SRT* factor (Eq. (36) and the solution of the system of Eq. (38)) results from the combined action of the trailer systems, which allows a body roll angle of the trailer model (**Figure 33**) [32]. In this figure, it can be seen how the stability factor varies according to the influence of some of the parameters of the developed model.

When the model considers all parameters, the *LLT* coefficient on the front axle is approximately 70% of the *LLT* coefficient on the rear axle [21]. Applying this concept, the SRT_{all} factor reduces to *0.3364 g*. Finally, the proposed model shows how the lateral offset of the cargo (d_1 = 0.1 m) influences the SRT_{off} factor: 2 cm of lateral offset corresponds to a loss of stability of around *0.01 g* a reduction similar to that reported by Winkler [20, 32].

Additionally, the proposed model shows how a change in the lateral separation between the springs (b) influences the *SRT* factor. Some *LCVs* with tanker trailers have a greater lateral separation between the springs, which leads to a decrease in the roll angle and thus an increase in the *SRT* factor: 1 cm of lateral separation between the springs corresponds to a gain or loss of stability of around *0.001 g*, as shown in **Figure 33b** [32].

This model also allows the determination of the lateral (h_1) and vertical (h_2) *CG* location (**Figure 34**).

Finally, if we consider the recommended maximum lateral load transfer ratio for the rear axle of 0.6 [39, 40], and also include the recommended bank angle and longitudinal slope of the road [41, 42], we can calculate the *SRT* factor for a trailer model on downhill and uphill corners. **Table 6** shows a trailer model with different trailer/trailer angles (ψ) [32].

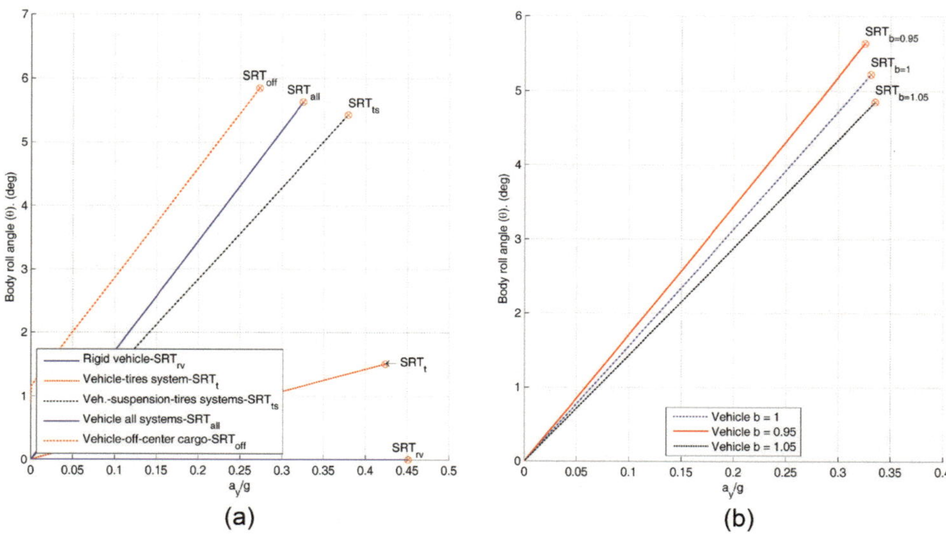

Figure 33.
(a) Roll angle of the trailer (θ). (b) Change in the SRT *factor.*

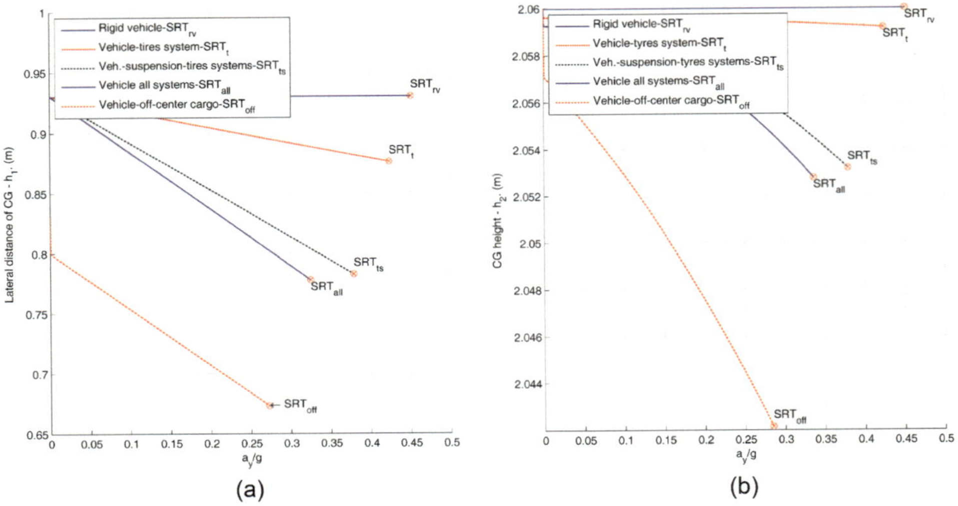

Figure 34.
(a) Lateral CG location. (b) Vertical CG location.

Slope angle (φ)-(%)		Uphill corners					Downhill corners			
Trailer/trailer angle (ψ)-(°)	**Bank angle (ϕ)-(%)**	**8**	**6**	**4**	**2**	**0**	**2**	**4**	**6**	**8**
0	0	0.240	0.223	0.214	0.208	0.202	0.198	0.194	0.190	0.186
0	2	0.261	0.245	0.235	0.228	0.223	0.218	0.214	0.210	0.206
0	4	0.283	0.265	0.256	0.249	0.244	0.239	0.234	0.230	0.226
0	6	0.305	0.286	0.277	0.270	0.265	0.260	0.256	0.252	0.248
0	8	0.327	0.308	0.299	0.292	0.285	0.281	0.276	0.272	0.268
0	10	0.350	0.330	0.320	0.313	0.307	0.303	0.298	0.294	0.290
10	0	0.241	0.223	0.214	0.207	0.202	0.197	0.193	0.189	0.185
10	2	0.262	0.245	0.234	0.227	0.222	0.217	0.213	0.209	0.205
10	4	0.283	0.265	0.256	0.249	0.243	0.239	0.233	0.229	0.225
10	6	0.306	0.286	0.276	0.270	0.264	0.259	0.255	0.251	0.247
10	8	0.328	0.308	0.298	0.291	0.285	0.280	0.276	0.271	0.267
10	10	0.351	0.330	0.319	0.312	0.307	0.302	0.297	0.293	0.288
20	0	0.241	0.222	0.212	0.206	0.200	0.195	0.191	0.187	0.182
20	2	0.262	0.244	0.233	0.226	0.220	0.216	0.211	0.207	0.203
20	4	0.283	0.264	0.254	0.247	0.242	0.236	0.232	0.227	0.224
20	6	0.306	0.285	0.275	0.268	0.262	0.257	0.253	0.249	0.245
20	8	0.328	0.307	0.297	0.290	0.283	0.278	0.274	0.270	0.265
20	10	0.351	0.329	0.318	0.311	0.305	0.300	0.295	0.291	0.286

Table 6.
Static rollover threshold (SRT) of the trailer model with trailer/trailer angle.

In the worst-case scenario, the trailer model, for a downhill corner with a bank angle of 0%, the longitudinal slope of the road of 8%, and a trailer/trailer angle of 20° can reduce the *SRT* factor of the model by 59.6%, using *0.4511 g* as a reference [32].

An analysis of **Table 6** leads to the following conclusions for the critical conditions of the trailer:

- a 1% bank angle corresponds to gain in the stability of around $0.01\,g$;
- when the trailer is in downhill corners, a 1% slope angle corresponds to a loss of stability of around $0.0021\,g$;
- the trailer/trailer angle is inversely proportional to the *SRT* factor since when the trailer makes a horizontal curve with a small radio, and the trailer/trailer angle and inertial force are large, the *SRT* factor is lower.

4. Conclusions

This study demonstrates that the longitudinal characteristics of a trailer model have an essential influence on the *SRT* factor calculation. In this case, the *SRT* factor is approximately 38% lower than the previously reported standard value. This value is very close to that reported by Winkler [20] (i.e. 40%), which suggests that the proposed model provides consistent results [32].

This model also shows that the change in the lateral separation between the springs (b) plays an important role, and thus it should be considered in the design and construction of trailers. Greater lateral separation between the springs will increase the trailer model stability [32].

We also found that the parameters of the road, such as the bank angle and the longitudinal slope angle, can affect vehicle stability. This situation is closer to the actual problem: when the road is not planar, the lateral and the longitudinal load transfer play an important role in reducing the stability. On the other hand, this provides a very important warning, because some simplifications carried out when estimating the *SRT* factor can lead to a considerably higher stability value. This is a point of concern, leading to the perception that our roads are safer than they really are [32].

Author details

Gonzalo Guillermo Moreno Contreras[1,2]*, Rodrigo de Souza Vieira[2] and Daniel Martins[2]

1 Mechanical Engineering Department, University of Pamplona, Colombia

2 Mechanical Engineering Department, Federal University of Santa Catarina (UFSC), Brazil

*Address all correspondence to: gmoren@unipamplona.edu.co; gmoren@hotmail.com

References

[1] Rempel MR. Improving the dynamic performance of multiply-articulated vehicles [Master's thesis]. Vancouver, Canada: The University of British Columbia; 2001

[2] Melo RP. Avaliação da estabilidade lateral de CVCs [Master's thesis]. Brazil: Pontifical University Catholic of Parana; 2004

[3] Ledesma R, Shih S. Heavy and medium duty vehicle suspension-related performance issues and effective analytical model for system design guide. International Truck & Bus Meeting & Exposition. SAE International; 1999. DOI: 10.4271/1999-01-3781

[4] Rill G. Road Vehicle Dynamics: Fundamentals and Modeling. Boca Ratón, Florida: CRC Press; 2011. ISBN: 978-1-4398-3898-3

[5] Crossley FRE. A contribution to Grübler's theory in number synthesis of plane mechanisms. ASME Journal of Engineering Industry. 1964;**86**(2):1-8

[6] Kutzbach K. Mechanische leitungsverzweigung, ihre gesetze und anwendungen. Maschinenbau Betrieb. 1929;**8**(8):710-716

[7] Tsai LW. Mechanism Design: Enumeration of Kinematic Structures According to Function. Boca Ratón, Florida: CRC press; 2001. ISBN: 0849309018

[8] Jazar RN. Vehicle Dynamics: Theory and Application. New York: Springer; 2014. ISBN: 978-1-4614-8544-5

[9] Pacejka H. Tire and Vehicle Dynamics. Netherlands: Elsevier Ltd; 2012. ISBN: 9780080970165

[10] Smith ND. Understanding parameters influencing tire modeling. Formula SAE Platform - Department of Mechanical Engineering - Colorado State University; 2004

[11] Erthal J. Modelo cinestático para análise de rolagem em veículos [PhD thesis]. Florianópolis, Brazil: Universidade Federal de Santa Catarina; 2010

[12] Lee U. A study on a method for predicting the vehicle controllability and stability using the screw axis theory [PhD thesis]. Seoul, South Korea: Hanyang University; 2001

[13] Mejia L, Simas H, Martins D. Force capability maximization of a 3rrr symmetric parallel manipulator by topology optimization. In: 22nd International Congress of Mechanical Engineering (COBEM 2013). Ribeirão Preto, SP, Brazil: ABCM - Associação Brasileira de Engenharia e Ciências Mecânicas; 2013

[14] Taylor RK, Bashford LL, Schrock MD. Method for measuring vertical tire stiffness. American Society of Agricultural Engineers. 2000;**42**(6):1415-1419

[15] Rill G et al. Leaf spring modelling for real time applications. In: 18th IAVSD-Symposium. Atsugi-Japan.: IAVSD-The International Association for Vehicle System Dynamics; 2003

[16] Moreno G, Frantz JC, Nicolazzi LC, Vieira RS, Martins D. Stiffness and deformation of mechanisms with locally flexible bodies: A general method using expanded passive joints. In: International Symposium on Advances in Robot Kinematics. Cham: Springer; 2018. pp. 285-292

[17] Dhoshi NP, Ingole NK, Gulhane UD. Analysis and modification of leaf spring of tractor trailer using analytical and finite element method. International Journal of Modern Engineering Research. 2011;**1**(2):719-722

[18] Saf-Holland. About Fifth Wheels. Germany: SAF-HOLLAND – Verkehrstechnik GmbH; 2006

[19] Kurdi O, Rahman RA, Samin PM. Optimization of heavy duty truck chassis design by considering torsional stiffness and mass of the structure. Applied Mechanics and Materials. 2014; **554**:459-463

[20] Winkler C. Rollover of Heavy Commercial Vehicles. SAE RR-004. Warrendale: Society of Automotive Engineers; 2000. p. 74. ISBN: 978-0-7680-0626-1; http://books.sae.org/rr-004/

[21] Kamnik R, Boettiger F, Hunt K. Roll dynamics and lateral load transfer estimation in articulated heavy freight vehicles. In: Proceedings of the Institution of Mechanical Engineers: Journal Automobile Engineering. 2003; **217**(11):985-997

[22] Davies TH. Mechanical networks—I. Passivity and redundancy. Mechanism and Machine Theory. 1983;**18**(2):95-101

[23] Davies TH. Mechanical networks—III. Wrenches on circuit screws. Mechanism and Machine Theory. 1983; **18**(2):107-112

[24] Davies TH. The 1887 committee meets again. Subject: Freedom and constraint. In: Ball 2000 Conference. Cambridge, UK: Cambridge University Press, Trinity College; 2000. p. 56

[25] Moreno GG, Nicolazzi L, Vieira RS, Martins D. Three-dimensional analysis of the rollover risk of heavy vehicles using Davies method. In: 14th World Congress in Mechanical and Machine Science (IFToMM2015), Taipei, Taiwan; 2015. DOI: 10.6567/IFToMM.14TH.WC.PS4.006

[26] Moreno G, Barreto RLP, Nicolazzi L, Vieira R, Martins D. Three-dimensional analysis of vehicles stability using graph theory. In: Graph-Based Modelling in Engineering. Switzerland: Springer International Publishing; 2016. DOI: 10.1007/978-3-319-39020-8_9. ISBN: 978-3-319-39018-5

[27] Moreno G, Nicolazzi L, Vieira RS, Martins D. Modeling and analysis of solid axle suspension and its impact on the heavy vehicles stability. In: Congresso Nacional de Engenharia Mecânica (CONEM2016); Fortaleza, Brazil; 2016. DOI: 10.13140/RG.2.2.32141.33766

[28] Moreno G, Nicolazzi LC, Vieira RDS, Martins D. Suspension and tyres: Stability of heavy vehicles. International Journal of Heavy Vehicle Systems. 2017;**24**(4):305-326. DOI: 10.1504/IJHVS.2017.087221

[29] Tsai LW. Robot Analysis—The Mechanism of Serial and Parallel Manipulators. New York: John Wiley & Sons; 1999. ISBN: 0-471-32593-7

[30] Davies TH. Couplings, coupling network and their graphs. Mechanism and Machine Theory. 1995;**30**(7): 1001-1012

[31] Heyman J. Basic Structural Theory. New York: Cambridge University Press; 2008. ISBN 13: 978-0-511-39692-2

[32] Moreno G, Nicolazzi LC, Vieira RDS, Martins D. Stability of long combination vehicles. International Journal of Heavy Vehicle Systems. 2018; **25**(1):113-131. DOI: 10.1504/IJHVS.2018.089897

[33] Cabral JC. Randon e Wabco Desenvolvem Sistema Electrónico Anti-Tombamento Para Bitrens—Agência Intelog Notícias. 2008. Available from: http://www.newslog.com.br [Accessed: 19 August 2008]

[34] Lui P, Rakheja S, Ahmed A. Detection of dynamic roll instability of heavy vehicles for open-loop rollover control. In: 1997 International Truck and

Bus Meeting. SAE Special Publications 1308 - SAE Paper No. 973263; Cleveland, Ohio; 1997. pp. 105-112

[35] Gillespie TD. Fundamentals of Vehicle Dynamics. Warrendale, PA: SAE International; 1992. ISBN: 1560911999

[36] Michelin Ltd. Michelin XZA Tire. Greenville, SC: Michelin North America, Inc.; 2013

[37] Harwood DW, Torbic DJ, Richard KR, Glauz WD, Elefteriadou L. Review of Truck Characteristics as Factors in Roadway Design. Transportation Research Board. Washington DC: National Cooperative Highway Research Program; 2003. ISBN: 0-309-08779-1

[38] Ervin RD, Guy Y. The influence of weights and dimensions on the stability and control of heavy duty trucks in Canada. UMTRI - The University of Michigan Transportation Research Institute, Final Report UMTRI-86-35/III; 1986

[39] Walker HK, Pearson JR. Recommended regulatory principles for interprovincial heavy vehicle weights and dimensions. Tech. Rep., CCMTA/RTAC Vehicle Weights and Dimensions Study Implementation Committee Report; 1987

[40] Woodrooffe J, Sweatman P, Arbor A, Middleton D, James R, Billing JR. National Cooperative Highway Research Program—NCHRP. Report 671. Review of Canadian Experience with the Regulation of Large Commercial Motor Vehicles. Washington, D.C.: National Academy of Sciences; 2010. ISBN: 978-0-309-15518-2

[41] AASHTO. A Policy on Geometric Design of Highways and Streets. Tech. Rep. 4th ed. Washington, D.C.: AASHTO; 2001. ISBN: 1-56051-156-7

[42] AASHATO. Recommendation for AASHTO Superelevation Design. Washington, D.C.: Design Quality Assurance Bureau, NYSDOT; 2003